UB Siegen
61 TKO2687

D1660704

CLUSTER DISSECTION AND ANALYSIS
Theory, FORTRAN Programs, Examples

## ELLIS HORWOOD SERIES IN COMPUTERS AND THEIR APPLICATIONS

*Series Editor:* Brian Meek, Director of the Computer Unit, Queen Elizabeth College, University of London

|  |  |
|---|---|
| Atherton, R. | **Structured Programming with COMAL** |
| Berry, R. E. | **Programming Language Translation** |
| Brailsford, D. F. and Walker, A. N. | **Introductory ALGOL 68 Programming** |
| Bull, G. M. | **The Dartmouth Time Sharing System** |
| Burns, A. | **New Information Technology** |
| Burns, A. | **The Microchip: Appropriate or Inappropriate Technology** |
| Chivers, I. D. and Clark, M. W. | **Interactive FORTRAN 77: A Hands-on Approach** |
| Cope, T. | **Computing using BASIC: An Interactive Approach** |
| Dahlstrand, I. | **Software Portability and Standards** |
| Davie, A. J. T. and Morrison, R. | **Recursive Descent Compiling** |
| Deasington, R. J. | **A Practical Guide to Computer Communications and Networking Second Edition** |
| Deasington, R. J. | **X.25 Explained** |
| Ennals, R. | **Logic Programmers and Logic Programming** |
| Fossum, E. | **Computerization of Working Life** |
| Gray, P. M. D. | **Logic, Algebra and Databases** |
| Harland D. | **Concurrent Programming** |
| Harland, D. M. | **Polymorphic Programming Languages** |
| Hill, I. D. and Meek, B. L. | **Programming Language Standardisation** |
| Hutchison, D. | **Fundamentals of Computer Logic** |
| McKenzie, J., Elton, L. and Lewis, R. | **Interactive Computer Graphics in Science Teaching** |
| Matthews, J. | **FORTH** |
| Meek, B. L., Fairthorne, S. and Moore, L. | **Using Computers, 2nd Edition** |
| Meek, B. L., Heath, P. and Rushby, N. | **Guide to Good Programming Practice, 2nd Edition** |
| Millington, D. | **Systems Analysis and Design for Computer Application** |
| Moore, L. | **Foundations of Programming with PASCAL** |
| Paterson, A. | **Office Systems: Planning, Procurement and Implementation** |
| Pemberton, S. and Daniels, S. C. | **PASCAL Implementation** |
| Pesaran, H. M. and Slater, L. J. | **Dynamic Regression: Theory and Algorithms** |
| Peter, R. | **Recursive Functions in Computer Theory** |
| Ramsden, E. | **Microcomputers in Education 2** |
| Sharp, J. A. | **Data Flow Computing** |
| Smith, I. C. H. | **Microcomputers in Education** |
| Spath, H. | **Cluster Analysis Algorithms** |
| Spath, H. | **Cluster Dissection and Analysis** |
| Stratford-Collins, M. J. | **ADA: A Programmer's Conversion Course** |
| Teskey, F. N. | **Principles of Text Processing** |
| Turner, S. J. | **An Introduction to Compiler Design** |
| Young, S. J. | **An Introduction to ADA, Second (Revised) Edition** |
| Young, S. J. | **Real Time Languages** |

## ELLIS HORWOOD BOOKS IN COMPUTING

*Series Editor:* A. J. Jones, Brunel University

|  |  |
|---|---|
| Atherton, R. | **Structured Programming with BBC BASIC** |
| Barrett, T. P. and Colwill, S. | **Winning Games on the Commodore 64** |
| Barrett, T. P. and Jones, A. J. | **Winning Games on the VIC-20** |
| Christensen, B. | **Beginning COMAL** |
| Cole, D. G. J. | **Getting Started on the ORIC-1** |
| Ennals, R. et al. | **Information Technology and the New Generation** |
| Ennals, R. | **Beginning micro-PROLOG, 2nd Revised Edition** |
| Goodyear, P. | **Commodor 64 LOGO** |
| Goodyear, P. | **LOGO: A Guide to Learning through Programming** |
| Hepburn, P. | **Problem-Solving with micro-PROLOG** |
| Jones, A. J. and Carpenter, G. | **Mastering the Commodore 64** |
| Jones, A. J., Coley, E. A. and Cole, D. G. J. | **Mastering the VIC-20** |
| Matthews, T. and Smith, P. | **Winning Games on the ZX Spectrum** |
| Matthews, T. and Smith, P. | **Winning Strategy Games on the Commodore 64** |
| Moore, L. | **Mastering the ZX Spectrum** |
| Narayanan, A. | **Beginning LISP** |
| Simon and Matthews, J. | **Mastering the Electron** |
| Whiddett, R. J. | **Getting to Grips with UNIX** |

# CLUSTER DISSECTION AND ANALYSIS

## Theory, FORTRAN Programs, Examples

HELMUTH SPÄTH
Professor of Mathematics
University of Oldenburg, Oldenburg, West Germany

*Translator:*
JOHANNES GOLDSCHMIDT
Department of Community Medicine
St. Thomas' Hospital Medical School, London

ELLIS HORWOOD LIMITED
Publishers · Chichester

Halsted Press: a division of
JOHN WILEY & SONS
Chichester · New York · Ontario · Brisbane

This English Edition first published in 1985 by
**ELLIS HORWOOD LIMITED**
Market Cross House, Cooper Street, Chichester, West Sussex, PO19 1EB, England

*The publisher's colophon is reproduced from James Gillison's drawing of the ancient Market Cross, Chichester.*

**Distributors**
*Australia, New Zealand, South-east Asia:*
Jacaranda-Wiley Ltd., Jacaranda Press,
JOHN WILEY & SONS INC.
GPO Box 859, Brisbane, Queensland 4001, Australia
*Canada:*
JOHN WILEY & SONS CANADA LIMITED
22 Worcester Road, Rexdale, Ontario, Canada
*Europe, Africa:*
JOHN WILEY & SONS LIMITED
Baffins Lane, Chichester, West Sussex, England
*North and South America and the rest of the world:*
Halsted Press: a division of
JOHN WILEY & SONS
605 Third Avenue, New York, NY 10016, USA

This English edition is translated from the author's original German edition
*Cluster-Formation-und-Analyse Theorie, FORTRAN Programme, Beispiele*
published in 1983 by R. Oldenbourg Verlag GmbH, Munich, © the copyright holders



**British Library Cataloguing in Publication Data**
Späth, Helmuth
Cluster dissection and analysis: theory FORTRAN programs, examples. –
(Ellis Horwood series in computers and their applications)
1. Cluster analysis – Computer programs
I. Title II. Cluster Formation und Analyse. *English*
519.5'3 QA278

**Library of Congress Card No.** 84-25240

ISBN 0-85312-736-0 (Ellis Horwood Limited)
ISBN 0-470-20129-0 (Halsted Press)

Printed in Great Britain by R.J. Acford, Chichester.

# Table of Contents

# Preface

This book was written after more than a dozen years of personal experience in practical applications and research. It follows a monograph written eight years earlier by the author, numerous seminars for practitioners, and lectures on the subject in Oldenburg and Rio de Janeiro. The intention was to write a typical textbook of applied mathematics which, after careful study with the use of a computer, would be equally useful in teaching, research, and practical application.

However, the book is written not only for mathematicians. Since there are many disciplines where large and complex masses of data arise, either inevitably or by design, the book is addressed to all who find themselves in this situation, and who have an elementary knowledge of matrix algebra and of FORTRAN.

In view of this wide circle of interested people, the book hopes to contribute towards a re-orientation away from the convenient but frequently incompetent use of large software packages and towards a competent and flexible use of the smaller programs shown in this book. These can be understood in their full detail once the theory has been studied.

Part I comprises a self-contained theoretical exposition of so-called partitioning cluster methods with an objective function (e.g. minimum variance criterion). In the author's experience and opinion, these various methods are the most important when dealing with large quantities of data. Because the stochastic models frequently found in the literature rely on assumptions which are rarely fulfilled or rarely verifiable with the data sets found in practice, the methods contained in this book will be presented from a geometrical point of view as combinatorial optimization problems of descriptive statistics. The approximation methods used are simple but efficient, especially as some recent findings of numerical analysis are used which are unlikely to have been taken into account in other people's programs.

Part II contains details of the algorithms and their implementation, which are necessary for an understanding and competent application of the listed machine independent FORTRAN IV subroutines.

Part III contains sample main programs and many worked examples. In actual applications it would be necessary to attempt an interpretation of the output, but this has been deliberately omitted here. Since the output from the examples is presented geometrically, it is still possible to give important hints on the choice of method, and on the procedures to be used for other tasks.

The Appendix includes a description of a magnetic tape written in standard format which includes all the programs quoted in the text, and which may be obtained. In addition, hints for the implementation in FORTRAN-80 on microcomputers are given.

At the end of each section is a number of problems to be worked through. In part I of the book these tend to be the usual type of exercises. In parts II and III, however, it will be necessary to use a computer after the mathematical groundwork is done. At times it will also be necessary to consult original publications.

The references are split into two parts. Part A is intended to contain all important monographs and survey articles about cluster analysis. Since there are by now over 3000 original publications (see A2, A4, A5, A10, A12, A17), we did not try to list and quote all of them individually. Part B is restricted to those references which have a direct bearing on the material presented here. This includes references to books and original papers in numerical analysis. In keeping with the textbook character of this book, we do not always explicitly refer to all works which have contributed to the formulation of certain statements, especially not to references in part A.

### *Preface to the English Edition*

Some printing errors have been corrected. Corrections to the programs were not necessary. Three figures have been added to the appendix.

### *Acknowledgement*

Many thanks to that most congenial translator, J. Goldschmidt, London.

### *Address of the Author*

Prof. Dr. Helmuth Späth,
Fachbereich Mathematik, Universität Oldenburg, Postfach 2503,
2900 Oldenburg, Federal Republic of Germany.

# 1 Introduction

Let us start with some examples where the cluster methods to be described have been successful or where they could be applied.

**Case 1 [B32]**
The textile industry wishes to determine new sizes for men's suits (e.g. normal men's sizes). A size is defined in terms of body height, chest, waist, inside and outside leg measurements. One size for all would be ideal from the manufacturing point of view, but could hardly be sold successfully, since it would be a bad fit for too many men. If a large number of sizes are manufactured – ideally made to measure – this is very expensive and creates inventory problems, but is ideal for the customer. As a compromise, one tries to find a limited number of sizes, say ten, but tries to choose these so that they are the best possible fit for the largest possible number of men. What does one do? One chooses a random sample of approximately 2000 normal men and measures on them the values of the variables mentioned above which define the size of the suits. The mean of each variable would determine the one standard size. In the case of ten sizes, one should be looking for ten 'best' means. For example, these could be defined so that the sum of variances for the groups of men defining each size is minimized. The task is to split the 2000 men into 10 groups so that this is the case. Later, this objective function will be referred to as the minimum variance criterion.

**Case 2 [B44]**
A mail order firm wants to develop a system of forecasting demand for a range of approximately 40 000 products whose composition changes. Experience shows that certain product groups have approximately the same seasonal demand pattern, while the demand for production in other groups differs. In order to use the recently obtained information on demand in the current catalogue season to forecast future demand, the 40 000 products from former catalogue seasons are split into categories according to the seasonal pattern. Time series are constructed showing the demand for each pattern for each of 18 previous 10 day periods. Assuming the continuation of past trends, this permits a forecasting of demand for a new product as soon as it has been assigned to one of the patterns. The problem lies in defining seasonal patterns within which the differences in the trend of demand are minimized.

**Case 3 [B52]**
A life style analysis is carried out to find the types of consumer who are particularly susceptible to certain kinds of advertising and certain products. A

representative sample of approximately 5000 men or women is given a questionnaire. In addition to age, household income, education, reading habits, town size, etc., the responses to so-called AIO statements (Activities, Interests, Opinions) are recorded. The interviewee is asked to record his attitude to questions such as 'Should a woman smoke in the street?' on an ordinal scale with values between 1 and $t$. Frequently, $t=7$. Assuming that answers reflect typical consumer behaviour, one is looking for consumer types with similar response patterns. Having defined these consumer types in terms of responses to AIO statements, one then calculates mean values for the other recorded characteristics, so that product type specific advertising can be arranged.

**Case 4**

In a department store, approximately 1000 customers are asked which of approximately 50 product types they have bought in one visit. For the appropriate positioning of the sales stands it is of interest whether certain combinations of products are frequently bought together.

**Case 5 [B8]**

A petrol company wants to assess the performance of its approximately 5000 filling stations. The filling stations are characterized in terms of location, size, length of street frontage, road type (motorway, A road, B road, main road, or side street), operating style (self service, attendant service, or mixed), number of pumps, car wash, and service facilities. The question is whether stations operating under similar conditions achieve similar sales, and which stations perform worse than expected and should perhaps be closed. Clearly, it is necessary to divide the stations into groups with similar characteristics, and to calculate, for each group, the average sale and the deviation of individual stations from this average.

**Case 6**

To find out which of 15 well-known politicians are perceived as similar, a representative sample of people is given all 105 possible pairs of these politicians, and asked to rank the pairs in terms of their similarity. By suitable aggregation of the rankings given by several people, a symmetric distance matrix is obtained whose values can be used to find groups of similar politicians, with as much difference as possible between the groups. This could, for example, be relevant in forming election campaign teams.

**Case 7**

An association of beef farmers wants a forecast of beef consumption in the coming year. The data available for the past 50 years include the actual beef consumption and three factors which, amongst others, are thought to determine it, namely beef price, per capita pork consumption, and per capita income. After adjusting for inflation, one could try to express beef consumption as a linear function of the three determining factors, to use independent forecasts of these three quantities to evaluate the function, and thus forecast the next year's consumption. However, since the past 50 years include times of war, and since a number of other factors also affected beef consumption, there is a plausible hypothesis suggesting that different linear relationships hold true for different groups of years, and that functions with different coefficients for each of these sets describe the underlying relationship better than a single function. Before the forecast is made, it would be necessary to assign the year of the forecast to one of these groups.

---

What is the common element of all these questions which occur not only in an economic or social science context, but also in almost all empirically based disciplines including biology, medicine, psychology, geology, archaeology, engineering and computer science?

In the framework of a specific objective, an important subtask is the formation or discovery of suitable or existing groups, whose number is either predetermined or to be discovered. These groups must be as homogeneous as possible internally while being as different from each other as possible. More precise definitions will follow below.

The objects to be grouped together may be characterized either by the values of s variables

$$X = (x_{ik}) \quad (i=1, \ldots, m, k=1, \ldots, s) \tag{1.1}$$

or by the distance values

$$t_{ih} = t_{hi} \quad (h=1, \ldots, i, i=2, \ldots, m). \tag{1.2}$$

The variables or distances should be chosen according to the purpose to be achieved by the grouping. Sometimes secondary characteristics and their values also play a part.

The number of objects m in the data matrix (1.1) tends to be large, and larger than the number of variables s. Sometimes, however, this is not so. In Cases 1 to 5 above the values of (m,s) are (2000,5), (40 000,18), (5000,300), (50,1000), and (5000,7) respectively. The variables may be quantitative (i.e. their range and properties are those of real numbers) as in Cases 1 and 2, or ordinal (i.e. their values can be given a rank order) as in Case 3, or binary or nominal (without order) as in Case 4. Finally, mixtures of all these may occur, as in Case 5.

Where distances (1.2) are given, the number of objects cannot be very large if only because of the quantity of data involved, as the number of distances between m items is $m(m-1)/2$. Nevertheless there are, even here, cases with large numbers of objects. Take Case 5, for example, where mixed data types make it impossible to define centres (analogous to the mean values of quantitative variables) or to measure homogeneity (analogous to the sum of squared deviations from the mean of quantitative variables). In such a case, it is possible to use the data matrix (1.1) to construct a suitable distance matrix (1.2). In Case 5 one would need to calculate and store about 12.5 million distances. Of course, there is the possibility of using a representative sample.

In any event, the groups sought are called classes or clusters. Where the aim is to create suitable classes whose number is arbitrary or suggested by the purpose of the classification, the term 'cluster dissection' is used [B32]. Where, however, one suspects the existence of an underlying structure (which might be the case in the other examples), and tries to discover the clusters and their number, the term 'cluster analysis' is appropriate. Where the use of cluster techniques does not bring to light an existing class structure, but imposes a division into classes which might even depend on the specific method used, the word 'analysis' should be avoided. However, the results may still be useful if the clusters imposed at least have certain properties. One such property might be that the distance from each element of a cluster to the centre of the cluster is no greater than the distance to the centres of the other clusters, i.e. that a so-called minimal distance partition is produced.

If we number the m objects and identify them by these numbers, that is, if we denote the set of objects as

$$M = \{1, \ldots, m\} \tag{1.3}$$

then cluster dissection or cluster analysis with n clusters means that we seek subsets $C_j \subset M$ with

$$\begin{aligned}
&C_1 \cup \ldots \cup C_n = M, \\
&C_i \cap C_k = \phi \ (i \neq k), \quad (\phi \text{ empty set}) \\
&m_j := |C_j| \geqslant 1 \ (j=1, \ldots, n), \\
&1 \leqslant n \leqslant m
\end{aligned} \tag{1.4}$$

Notation (1.4) signifies that each object is contained in exactly one cluster and that each class contains at least one object. We identify class $C_j$ with the set of numbers which denote its members.

$$C = (C_1, \ldots, C_n) \tag{1.5}$$

with properties (1.4) is called a partition of length n. Thus C1={2,3,7}, C2={1,4}, C3={5,6,9}, C4={8}, is a partition of length 4 of 9 objects numbered 1 to 9. P(n,M) denotes the set of all partitions C of M which have length n.

The number S(n,m) of elements of P(n,M) is given [A8] by

$$S(n,m) = \frac{1}{n!} \sum_{j=1}^{n} (-1)^{n-j} \binom{n}{j} j^m \tag{1.6}$$

For m = 20 and n = 4, S(n,m) is as large as 45 232 115 901, and for m = 100, n = 5, S(n,m) exceeds $10^{68}$.

A partitioning clustering algorithm or criterion is defined as an objective function D which associates with each partition in P(n,M) a non-negative real number, and thus allows a comparison between the partitions to be made. An optimum partition can then be defined as one which maximizes or minimizes the objective function.

In this book, we shall restrict ourselves to a thorough treatment of such partitioning criteria for cluster dissection and analysis. We shall omit the methods listed below, because they have been treated elsewhere [A1, A4, A5, A10, A11, A20, A22, A23] and because they do not appear appropriate for examples like the ones above:

– visual methods [B9] in which a data matrix with $s>2$ is reduced to two variables, and the resulting points on a plane are examined by eye. The reduction uses methods such as multidimensional scaling or principal components analysis [A4, A22], which involve loss of information.
– hierarchical methods [A1, A4, A13] which produce hierarchies of partitions, each of which is formed from the preceding one by amalgamation or splitting of classes. These methods provide no criteria for judging one partition with a given number of classes better than another, since only the hierarchy building process is optimized.
– methods for density estimating on [A4, A5], since we want to avoid making stochastic assumptions which often cannot be verified in the actual sets of data dealt with.
– methods based on graph theory [A4, A5] which tend to involve a deliberate loss of information before they can be applied, and which are unsuitable for large numbers of objects.
– methods which rely on a given ordering of the distances [A15, A16, A17], since these are also unsuitable for large numbers of objects.

In addition, this book does not deal with a closely related generalization from the 'hard' partitions defined above to 'fuzzy' partitions [A2, A17], in which the degree $u_{ij}$ to which i belongs to class j is determined at the same time. We are of the opinion that, for large m, such information cannot be utilized sensibly, and that the resulting partition tends to be transformed into a 'hard' one.

As the examples, which are typical of applications, show, the choice of objective function using stochastic methods [A4] is often not advisable, as the data cannot be shown to conform to the necessary assumptions. As a result, we shall only rely on elementary pieces of descriptive statistics, and on certain geometric generalizations. For example, a desirable property of an objective function might be that, for an optimal partition, the convex hulls of the cluster members do not intersect.

Thus, our task of determining optimum partitions relative to suitable objective functions is, because of the finite number (1.6) of possibilities, a combinatorial optimization problem of descriptive statistics.

The number (1.6) is, for realistic values of m and n, too large for enumeration. Since it is unlikely that an algorithm for a sensible objective function can be found which will guarantee an optimum partition within polynomially bounded computation time, approximative methods need to be used. These do provide, in an efficient manner, partitions which may differ depending on initial values, but which at least fulfil a necessary condition for optimum partitions. The situation is comparable with optimizing a continuous function of several variables, where one is concerned with numerical methods for finding stationary points at which the gradient is zero.

# Part I

# Theory and methods

# 2 The minimum variance criterion

## 2.1 Derivation as the principle of several 'best' means

In Chapters 2 to 5, the objects to be classified will always be identified with the sets of numeric values used to describe their quantitative characteristics. Thus, for each object i there is a vector

$$x_i \in \mathbb{R}^s \quad (i=1, \ldots, m) \tag{2.1}$$

which is identical to the $i^{th}$ row of the data matrix (1.1).

Consider a fixed partition $C = (C_1, \ldots, C_n)$ of $M = \{1, \ldots, m\}$ with length n, where $1 \leqslant n \leqslant m$. Take one of the clusters $C_j$. As is generally known for $s = 1$, it is easily possible to calculate a centre $\bar{x}_j$ which is defined such that the sum of squared differences between it and all the numbers $x_i$ (a measure of scatter) is minimized. $\bar{x}_j$ is the arithmetic mean. This is also true for $s > 1$. Let the sum of squared deviations from an unknown centre $y \in \mathbb{R}^s$ be given by the function

$$S(y) := \sum_{i \in C_j} \| x_i - y \|^2 = \sum_{i \in C_j} \sum_{k=1}^{s} (x_{ik} - y_k)^2, \tag{2.2}$$

then the vector

$$\bar{x}_j := \frac{1}{m_j} \sum_{i \in C_j} x_i \quad (m_j := |C_j|) \tag{2.3}$$

minimizes the function $S = S(y)$. In calculating the mean vector $\bar{x}_j$, the mean for each component is calculated separately. Formally, this follows from the necessary condition

$$\frac{\partial S}{\partial y} = -2 \sum_{i \in C_j} (x_i - y) = 0. \tag{2.4}$$

The mean is indeed a global minimum, since

$$\frac{\partial^2 S}{\partial y^2} = 2m_j I \quad \text{(where I is the identity matrix).}$$

(2.4) implies the relationship

$$\sum_{i \in C_j} (x_i - \bar{x}_j) = 0. \tag{2.5}$$

which will be needed later. The value

$$e_j := e(C_j) := \sum_{i \in C_j} \| x_i - \bar{x}_j \|^2 \tag{2.6}$$

of the sum of squared deviations from the mean vector may be regarded as a measure of the compactness of a cluster. It is plausible to define the arithmetic mean of these measures of compactness, that is

$$\frac{1}{n}(e_1 + \ldots + e_n)$$

as a measure of the goodness of a partition. Omitting the factor $1/n$ (which is constant for fixed n), we have

$$D(C) = \sum_{j=1}^{n} e(C_j) = \sum_{j=1}^{n} \sum_{i \in C_j} \| x_i - \bar{x}_j \|^2 \tag{2.7}$$

$$= \sum_{j=1}^{n} \sum_{i \in C_j} \sum_{k=1}^{s} (x_{ik} - \bar{x}_{jk})^2 .$$

This is the first of the objective functions which we are looking for. It is called the sum of squared deviations or variance criterion. Since variance is usually defined as $e_j/(m_j - 1)$ the name 'variance criterion' is somewhat misleading. Nevertheless, it is widely used.

A partition $C_o \in P(n,M)$ is called optimal relative to (2.7) when

$$D(C_0) = \min_{C \in P(n,M)} D(C). \tag{2.8}$$

We try to find such partitions $C_o$.

Clearly, the minimum variance, as described above, corresponds to the principle of finding the n best means, which was motivated by Case 1 in Chapter 1.

## 2.2 Derivation from the decomposition of total scatter (T=W+B)

In this section, we wish to derive the minimum variance criterion in a different way which will enable us to obtain intermediate results of use in later chapters.

We shall also be led to a theorem which holds true for the minimum variance criterion, but not for the other objective functions discussed.

First, let us recall some important properties of the trace function which is defined for square matrices $A \in \mathbb{R}(s,s)$. The trace is a mapping $\mathrm{tr}: \mathbb{R}(s,s) \rightarrow \mathbb{R}$, defined (see [B30] as

$$\mathrm{tr}\,A := \sum_{k=1}^{s} a_{kk}, \quad A = (a_{ik}). \tag{2.9}$$

For $A, B \in \mathbb{R}(s,s)$, $\beta \in \mathbb{R}$, $y \in \mathbb{R}^s$ we have

$$\begin{aligned}
&\mathrm{tr}\,(A + B) = \mathrm{tr}\,A + \mathrm{tr}\,B,\\
&\mathrm{tr}\,(\beta A) = \beta\,\mathrm{tr}\,A,\\
&\mathrm{tr}\,(AB) = \mathrm{tr}\,(BA),\\
&\mathrm{tr}\,A = \sum_{k=1}^{s} \lambda_k \quad \text{(where } \lambda_k \text{ are the eigenvalues of A)}\\
&\mathrm{tr}\,(yy^T) = y^T y = \|y\|^2.
\end{aligned} \tag{2.10}$$

The symmetric matrix

$$yy^T = (y_i\, y_k)_{(i,\, k=1,\, \ldots,\, s)} \tag{2.11}$$

is known as the dyadic product of y with itself.

For an element $C_j$ of a given partition we shall now derive a relationship which will be doubly useful later. For any $y \in \mathbb{R}^s$ we start with the trivial identity

$$x_i - y = (x_i - \bar{x}_j) + (\bar{x}_j - y)$$

For each side of the equation, we form the dyadic product with itself:

$$(x_i-y)(x_i-y)^T = (x_i-\bar{x}_j)(x_i-\bar{x}_j)^T + 2(x_i-\bar{x}_j)(\bar{x}_j-y)^T + (\bar{x}_j-y)(\bar{x}_j-y)^T.$$

Adding terms for all $i \in C_j$ on the left as well as on the right, and taking into account (2.5), we obtain

$$\sum_{i \in C_j} (x_i-y)(x_i-y)^T = \sum_{i \in C_j} (x_i-\bar{x}_j)(x_i-\bar{x}_j)^T + m_j(\bar{x}_j-y)(\bar{x}_j-y)^T. \tag{2.12}$$

Now let us replace $y = \bar{x}$ in (2.12), where

$$\bar{x} = \frac{1}{m} \sum_{i=1}^{m} x_i\,, \tag{2.13}$$

then we obtain, for each class $C_j$, the relationship

$$\sum_{i \in C_j} (x_i - \bar{x})(x_i - \bar{x})^T = W_j + B_j, \tag{2.14}$$

where

$$W_j = \sum_{i \in C_j} (x_i - \bar{x}_j)(x_i - \bar{x}_j)^T, \tag{2.15}$$

$$B_j = m_j(\bar{x}_j - \bar{x})(\bar{x}_j - \bar{x})^T. \tag{2.16}$$

By adding (2.14) for all n classes, we obtain

**Theorem 2.1**: Using the notation above, the following relationship holds:

$$T = W + B, \text{ where} \tag{2.17}$$

$$T = \sum_{i=1}^{m} (x_i - \bar{x})(x_i - \bar{x})^T, \tag{2.18}$$

$$W = W(C) = \sum_{j=1}^{n} W_j, \quad B = B(C) = \sum_{j=1}^{n} B_j. \tag{2.19}$$

All the matrices $W_j$, $B_j$, W, B, $T \in \mathbb{R}(s,s)$ are symmetric. Relationship (2.17) is true for every partition C, but whereas the right hand side is dependent on C, the left hand side is not. Since T describes the total scatter, W that within classes, and B that between classes, Theorem 2.1 is known as the theorem about the decomposition of scatter. $W_j/(m_j-1)$ is commonly known as the empirical covariance matrix, but there appears to be no uniform terminology for $W_j$ itself. We call $W_j$ the scatter matrix of the $j^{th}$ class. As a result of (2.19), the variance criterion can also be expressed in the form

$$\min_{C \in P(n,M)} \operatorname{tr} W \tag{2.20}$$

This notation disguises the dependence of W on C. Relationship (2.17) permits another, equivalent, expression about the variance criterion to be stated:

**Theorem 2.2**: For the objective functions

$$\min_{C \in P(n,M)} \sum_{j=1}^{n} \sum_{i \in C_j} \| x_i - \bar{x}_j \|^2$$

$$\text{and} \max_{C \in P(n,M)} \sum_{j=1}^{n} m_j \| \bar{x}_j - \bar{x} \|^2 . \tag{2.21}$$

there are identical optimal partitions.

The significance of this statement is that minimizing the variance criterion, that is, minimizing the sum of squared deviations of the class members from the class mean, is equivalent to maximizing the sum of squared deviations of the class means $\bar{x}_j$ from the overall mean $\bar{x}$, provided the sum is weighted by the number $m_j$ of members in each class. In other words, in using the minimum variance criterion, one is aiming for classes which are internally as homogeneous as possible, but externally as different from each other as possible. Thus the minimum variance criterion has one important property which is essential for practical applications.

The proof of theorem 2.2 is quite simple. Starting with relationship (2.17), taking the trace of both sides, and remembering (2.10), we obtain

$$\sum_{i=1}^{m} \| x_i - \bar{x} \|^2 = \sum_{j=1}^{n} \sum_{i \in C_j} \| x_i - \bar{x}_j \|^2 + \sum_{j=1}^{n} m_j \| \bar{x}_j - \bar{x} \|^2 .$$

Since the left hand side is independent of the partition C, minimizing the first term on the right hand side is equivalent to maximizing the second term.

It is worth mentioning that there is no relationship analogous to (2.17) involving the empirical covariance matrices $W_j/(m_j-1)$, nor indeed $W_j/m_j$, and their sums. This appears to be the reason why a different, similarly plausible, criterion such as

$$\min_{C \in P(n,M)} \sum_{j=1}^{n} \frac{1}{m_j} \sum_{i \in C_j} \| x_i - \bar{x}_j \|^2$$

is not generally considered.

### 2.3 A formulation which does not involve means

The relationship (2.12) can be exploited in a different way. If we let $y = x_k$, sum both sides over all $k \in C_j$, and divide both sides by $2m_j$, we obtain (using definition (2.15))

$$W_j = \frac{1}{2m_j} \sum_{k \in C_j} \sum_{i \in C_j} (x_i - x_k)(x_i - x_k)^T . \qquad (2.22)$$

However, (2.10) implies

$$\text{tr } W = \sum_{j=1}^{n} \text{tr } W_j$$

and in conjunction with (2.20) we arrive at the following theorem.

**Theorem 2.3**: The variance criterion may be written in the form

$$\min_{C \in P(n,M)} \sum_{j=1}^{n} \frac{1}{m_j} \sum_{k \in C_j} \sum_{\substack{i \in C_j \\ i > k}} \| x_i - x_k \|^2 .$$

This formulation does not involve means. In each class, the sum of squared deviations of the class members from the class mean can be expressed as the sum of squared differences between individual class members, weighted by the inverse of the class size. This, too, is a property of the variance criterion which will not be found in the case of other, similar, objective functions.

## 2.4 Invariance properties and the problem of scaling

In linear algebra, the Euclidean distance between two vectors $x, y \in \mathbb{R}^s$ is given by

$$d(x,y) = \| x - y \| = \sqrt{(x-y)^T (x-y)} \tag{2.23}$$

It is invariant with respect to rotations and translations. For $A \in \mathbb{R}(s,s)$ and $b \in \mathbb{R}^s$,

$$d(x,y) = d(Ax + b, Ay + b) \tag{2.24}$$

holds if and only if A is an orthogonal matrix, that is, if

$$A^T A = AA^T = I \tag{2.25}$$

Now consider an arbitrary transformation

$$x_i' = Bx_i + b \quad (i=1, ..., m) \tag{2.26}$$

of our given vectors by $B \in \mathbb{R}(s,s)$. Clearly,

$$\bar{x}_j' = B\bar{x}_j + b \quad (j=1, ..., n) \tag{2.27}$$

holds, and thus for the variance criterion,

$$\sum_{j=1}^{n} \sum_{i \in C_j} \| x_i - \bar{x}_j \|^2 = \sum_{j=1}^{n} \sum_{i \in C_j} \| x_i' - \bar{x}_j' \|^2 \tag{2.28}$$

holds for any partition $C \in P(n,M)$ if and only if $B^T B = I$, that is, if B represents an orthogonal transformation. Thus, it is immaterial whether an optimal partition is sought for the vectors $x_i$ or $x_j'$ so long as B in (2.26) is an orthogonal matrix. We say that the variance criterion is invariant with respect to orthogonal transformations and translations.

In practical applications, this invariance property is not particularly important. A more interesting question is whether the composition of an optimal partition would change if scales of measurement are changed. Referring, for example, to Case 5, the area of the filling station might be specified in square metres or hectares, and the length of street frontage in metres or centimetres. In general, a scale transformation is defined as a non-singular diagonal matrix $H \in \mathbb{R}(s,s)$ with

$$H = \operatorname{diag}(h_1, \ldots, h_s) \quad (h_k \neq 0 \text{ for } k=1, \ldots, s) . \tag{2.29}$$

Because of (2.24) and (2.25), a scale transformation which leaves the variance criterion unchanged must satisfy (2.25).
This leads to the following theorem.

**Theorem 2.4**: The variance criterion is invariant with respect to scale transformations if

$$h_k = \pm 1 \quad \text{for } k=1, \ldots, s . \tag{2.30}$$

Thus, the composition of partitions which are optimal with respect to the variance criterion is unaffected only if a scale transformation amounts to no more than a change of sign. The application of arbitrary scale transformations (2.29) and, even more, the application of general linear transformations, can change the composition of optimal partitions. This is illustrated by an example: Let $m=6$ and $s=2$. The two data matrices

$$X = \begin{pmatrix} 0 & 48 \\ 6 & 48 \\ 0 & 24 \\ 6 & 24 \\ 0 & 0 \\ 6 & 0 \end{pmatrix} \quad \text{and} \quad X' = \begin{pmatrix} 0 & 12 \\ 24 & 12 \\ 0 & 6 \\ 24 & 6 \\ 0 & 0 \\ 24 & 0 \end{pmatrix}$$

are related to each other by

$$X' = XH \quad \text{where} \quad H = \begin{pmatrix} 4 & 0 \\ 0 & 1/4 \end{pmatrix} .$$

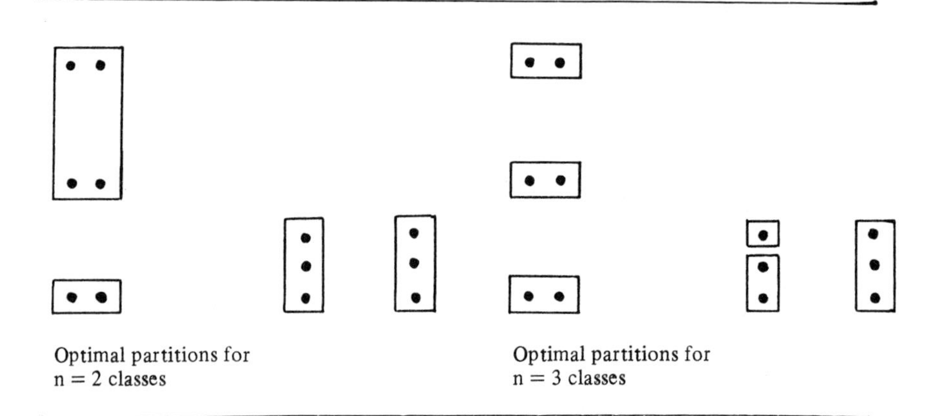


Fig. 2.1

Fig. 2.1 shows optimal partitions for n=2 and n=3 of X on the right and X′ on the left. The optimal partitions are not always unique. Not surprisingly, they differ in the two coordinate systems since H does not conform to (2.30). It is worth noting that for X′, n = 2 appears to be the 'natural', geometrically evident, number of classes, while for X, n=3 is the 'natural' number. Thus, if the minimum variance criterion is used, the results of a cluster analysis, including the determination of the 'natural' number of classes and the composition of the clusters, are all dependent on the units of measurement in which the data matrix is expressed. For practical applications, this is unfortunate.

To hide the problem a so-called standardization or z-transformation is often carried out before the variance criterion is applied:

$$x'_{ik} = (x_{ik} - \bar{x}_{.k})/s_k \qquad (i=1, \dots, m) \tag{2.31}$$

with

$$s_k^2 = \sum_{i=1}^{m} (x_{ik} - \bar{x}_{.k})^2 \qquad (k=1, \dots, s) \tag{2.32}$$

Here, $\bar{x}_{.k}$ is the mean of the $k^{th}$ column of the original data matrix $X = (x_{ik})$. The transformed matrix has the property

$$\bar{x}'_{.k} = 0, \quad s_k'^2 = 1 \qquad (k=1, \dots, s) \tag{2.33}$$

That is, all columns have mean zero and sum of squared deviations 1.

One should be aware that this type of scaling is just as arbitrary as any other, and has no special advantages. In Chapters 3 to 5 we shall deal with objective functions which are invariant with respect to changes in scale.

## 2.5 Up- and downdating formulae for calculating mean vectors and scatter matrices

In this section, we derive a formulae which are used here as well as in Chapters 3 and 4. We ask how the mean vector $\bar{x}_p$ defined in (2.3) and the scatter matrix $W_p$ defined in (2.15) change when an object i is either added or removed.

First, let

$$C_q := C_p \cup \{i\}, \quad i \notin C_p. \tag{2.34}$$

Then, clearly,

$$\bar{x}_q = \frac{1}{m_p + 1} (m_p \bar{x}_p + x_i). \tag{2.35}$$

For arbitrary $k \in M$, we have

$$x_k - \bar{x}_q = (x_k - \bar{x}_p) + (\bar{x}_p - \bar{x}_q) = (x_k - \bar{x}_p) + \frac{1}{m_p + 1} (\bar{x}_p - x_i).$$

If we form the dyadic product for both sides, we get

$$(x_k - \bar{x}_q)(x_k - \bar{x}_q)^T = (x_k - \bar{x}_p)(x_k - \bar{x}_p)^T$$
$$+ \frac{2}{m_p + 1}(x_k - \bar{x}_p)(\bar{x}_p - x_i)^T$$
$$+ \frac{1}{(m_p+1)^2}(\bar{x}_p - x_i)(\bar{x}_p - x_i)^T.$$

If we now sum both sides over all $k \in C_p$ or $k \in C_q$, respectively, and split the summation of the right hand side into $k \in C_p$ and $k = i$, we get, using (2.5),

$$W_q = W_p + \frac{m_p}{m_p + 1}(x_i - \bar{x}_p)(x_i - \bar{x}_p)^T \tag{2.36}$$

Taking the trace, we get

$$e_q = e_p + \frac{m_p}{m_p + 1} \| x_i - \bar{x}_p \|^2. \tag{2.37}$$

Analogously,

$$C_q := C_p \setminus \{i\}, \quad i \in C_p, \quad m_p > 1 \tag{2.38}$$

leads to formulae

$$\bar{x}_q = \frac{1}{m_p - 1}(m_p \bar{x}_p - x_i), \tag{2.39}$$

$$W_q = W_p - \frac{m_p}{m_p - 1}(x_i - \bar{x}_p)(x_i - \bar{x}_p)^T, \tag{2.40}$$

$$e_q = e_p - \frac{m_p}{m_p - 1} \| x_i - \bar{x}_p \|^2. \tag{2.41}$$

In both cases (2.34) and (2.38), new mean vectors, new scatter matrices, and new sums of squared deviations can be obtained in a very simple way, namely as a simple function of the previous value and of the difference $x_i - \bar{x}_p$.

### 2.6 Monotonicity

It is reasonable to expect the optimal partitions chosen by a criterion such as the minimum variance criterion to have the following property. As the length n of the partitions, that is, the number of classes, is increased, the values of the objective function for optimal partitions decrease. In fact, we have the following theorem.

**Theorem 2.5:** Let $1 \leqslant n \leqslant m$, and let $C = (C_1, \ldots, C_n)$ and $C' = (C'_1 \ldots, C'_{n+1})$ be partitions of length n and n+1, respectively, which are optimal according to the variance criterion. Then

$$D(C') \leqslant D(C) \ . \tag{2.42}$$

*Proof:* Since $D(C) \geqslant 0$ irrespective of the length of a partition, there is no need for proof when $n = m - 1$, since $D(C') = 0$ for $C' \in P(m,M)$ in that case. Hence, we may assume that $n+1 < m$. But then, each optimal partition C of length n contains at least one cluster $C_p$ with $m_p > 1$. Let $i \in C_p$ and define $C_q := C_p \setminus \{i\}$. From (2.41) we have

$$e_q = e_p - \frac{m_p}{m_p - 1} \| x_i - \bar{x}_p \|^2 \leqslant e_p \tag{2.43}$$

and

$$e_q = e_p \Longleftrightarrow x_i = \bar{x}_p. \tag{2.44}$$

Now consider the partition

$$C'' = (C_1, \ldots, C_{p-1}, C_q, C_{p+1}, \ldots, C_n, C_{n+1})$$

with length n+1, where $C_{n+1} := \{i\}$. $C''$ is not necessarily optimal. Using $e_{n+1} = 0$, (2.43), and the optimality of $C'$, we have

$$D(C') \leqslant D(C'') \leqslant D(C).$$

Furthermore, (2.44) implies that increasing the number of classes actually reduces the value of the objective function unless $x_i$ happens to be the mean vector in its class.

The property we have just proved is referred to as the monotonicity of the minimum variance criterion. It is also worth investigating for other objective functions.

## 2.7 Minimal distance partitions and minimal distance method

A partition $C = (C_1, \ldots, C_n)$ is called a minimal distance partition with respect to tr W if each object i (or $x_i$) has a minimum Euclidean distance from the mean vector $\bar{x}_j$ of the class $C_j$ to which it belongs. In other words, the squared deviation of any object from its own class mean is smaller than (or equal to) the squared deviations from all other class means, that is,

$$i \in C_j \Longleftrightarrow \| x_i - \bar{x}_j \| = \min_{k=1,\ldots,n} \| x_i - \bar{x}_k \| \, . \tag{2.45}$$

When the minimum distance occurs several times, that is, when j is not unique, we examine all these possible minimal distance partitions in turn. Then we choose that one as representative for which, in the chosen numbering of the classes, the minimum occurs for the smallest value of the index k.

Geometry suggests that for tr W an optimal partition is a minimal distance partition. This will be proved later. Conversely, however, the minimal distance property is not a sufficient condition for an optimal partition, as will become clear from the algorithm that follows and from the examples in section 11.4. The algorithm for generating minimal distance partitions iteratively has three steps, the second and third of which are repeated a finite number of times.

MINIMAL DISTANCE METHOD

*Step 1:* Set the operation count $t = 0$. Choose $ITMAX > 0$, and choose a starting partition $C_0 = (C_1^{(0)}, \ldots, C_n^{(0)})$ which is feasible in the sense of (1.4). The choice may be made randomly, for example.

*Step 2:* for the $t^{th}$ iteration calculate the mean vectors of $C^{(t)}$ according to

$$\bar{x}_j^{(t)} = \frac{1}{m_j^{(t)}} \sum_{i \in C_j^{(t)}} x_i \quad (j=1, \ldots, n). \tag{2.46}$$

*Step 3:* Calculate a minimal distance partition $C^{(t+1)} = (C_1^{(t+1)}, \ldots, C_n^{(t+1)})$, using

$$C_j^{(t+1)} := \{ i : \| x_i - \bar{x}_j \| = \min_{k=1,\ldots,n} \| x_i - \bar{x}_k \| \}, \tag{2.47}$$

By convention, take j to be the smallest index k for which the minimum is attained.

If $D(C^{(t+1)}) < D(C^{(t)})$, let $t = t + 1$ and return to step 2.

If $D(C^{(t+1)}) = D(C^{(t)})$, let $= t + 1$ as well, but go to step 2 only if $t \leqslant$ ITMAX. Otherwise, stop. The result is called an end partition of the minimal distance method.

First, we must show that the algorithm described does indeed produce progressively better approximations, that is, we must prove the following theorem.

**Theorem 2.6**: The iterations of the minimal distance algorithm do not increase tr W.

*Proof:* (see [A22]):

$$D(C^{(t)}) = \sum_{j=1}^{n} \sum_{i \in C_j^{(t)}} \| x_i - \bar{x}_j^{(t)} \|^2$$

$$\geqslant \sum_{j=1}^{n} \sum_{i \in C_j^{(t)}} \min_{k=1,\ldots,n} \| x_i - \bar{x}_k^{(t)} \|^2$$

$$= \sum_{j=1}^{n} \sum_{i \in C_j^{(t+1)}} \| x_i - \bar{x}_j^{(t)} \|^2$$

$$\geqslant \sum_{j=1}^{n} \sum_{i \in C_j^{(t+1)}} \| x_i - \bar{x}_j^{(t+1)} \|^2$$

$$= D(C^{(t+1)}).$$

This makes use of the minimal distance property and of the definition of the new partition according to (2.47), and then of the minimality property of the mean vector according to (2.2) and (2.3).

Because of Theorem 2.6, the method can only progress as follows: First, there is a strictly monotonic reduction in tr W. From iteration $t'$ on (where $t'$ may possibly equal zero) until iteration $t' + t''$ with $t'' \geqslant 1$,

$$D(C^{(t')}) = D(C^{(t'+1)}) = \ldots = D(C^{(t'+t'')}). \tag{2.48}$$

holds. If $t''$ is finite, there could follow

$$D(C^{(t'+t''+1)}) < D(C^{(t'+t'')})$$

and a continuation of the iterative process weight be possible. Because of the test for $t \leqslant$ ITMAX, the iterations might be stopped too soon for this to be detected. If ITMAX is chosen sufficiently large, this rarely happens in practice.

If $t''$ is infinite, there are two possibilities: Either $C^{(t')} = C^{(t'+1)} = \ldots$ for all following iterations, or the same end partitions are repeated at regular or irregular intervals, because the total number of partitions is finite. Both possibilities are handled by the test for $t \leqslant$ ITMAX in such a way that the iterations are not repeated in a senseless manner. However, the algorithm does not detect which of the two possibilities has occurred.

We must emphasize one weakness of the minimal distance method which shows up not only in theory but unfortunately also in practice. It may well happen that for one or several classes $C_j^{(t+1)} = \phi$, that is, that in the assignment of objects to classes one or more empty classes are generated. Because of (2.46) these remain empty. Thus, the method may produce end partitions which are not feasible according to (1.4), because they are too short. These are only acceptable if it is not necessary for the number of classes to remain constant. The partitions do have minimal distance properties.

## 2.8 The exchange method and a necessary condition for partitions to be optimal

The exchange algorithm is another method for finding minimal distance partitions. It will later be shown empirically shown to require less computer time, and will frequently lead to smaller values of the objective function.

EXCHANGE METHOD

*Step 1:* Choose an $m_0 \geqslant 1$. When using the minimum variance criterion. $m_0$ is usually chosen equal to 1. Set the iteration count $t = 0$. Choose ITMAX $> 0$ and an initial partition conforming to (1.4). If $m_0 > 1$, then the initial partition should also fulfil $m_j \geqslant m_0$ $(j=1, \ldots, n)$. Calculate the mean vectors $\bar{x}_j$ and the sum of squared deviations $e_j$ $(j=1, \ldots, n)$. Then choose an arbitrary starting object $i \in M$, for example $i = 1$.

*Step 2:* Let $i \in C_p$. If $m_p = m_0$, proceeed to step 3. Otherwise, try moving i into $C_j$ $(j=1, \ldots, n, j \neq p)$, and calculate $e(C_p \backslash \{i\})$ and $e(C_p \cup \{i\})$ for all $j=1, \ldots, n, j \neq p$. Such a move results in a change in the value of the objective function tr W. Using (2.37), (2.41) and the notation

$$C^{(j)} = (C_1 \ldots, C_p \setminus \{i\}, \ldots, C_j \cup \{i\}, \ldots, C_n)$$

the change amounts to

$$f_j := D(C) - D(C^{(j)}) \tag{2.49}$$

$$= e_p + e_j - e(C_p \setminus \{i\}) - e(C_j \cup \{i\})$$

$$= \frac{m_p}{m_p - 1} \| x_i - \bar{x}_p \|^2 - \frac{m_j}{m_j + 1} \| x_i - \bar{x}_j \|^2 .$$

If $f_j \leqslant 0$ for all $j \neq p$, then the attempt to exchange object i is unsuccessful in terms of reducing the value of the objective function. In this case, proceed to step 3. If, however, $f_j > 0$ for at least one $j \neq p$, and if therefore the value of the objective function can be reduced, then we want the reduction to be the largest possible. Hence, q is determined so that

$$f_q = \max_{j \neq p,\, f_j > 0} f_j . \tag{2.50}$$

If several indices q give the same maximum, choose the smallest one. Then perform the exchange $C_p := C_p \backslash \{i\}$ and $C_q := C_q \cup \{i\}$, and calculate new mean vectors and sums of squared deviations using formulae (2.39) and (2.35) or (2.41) and (2.37). Now proceed to

*Step 3:* Take the next object, that is, set $i := i + 1$. If $i > m$, restart with $i = 1$. In this case, set $t := t + 1$ and begin a new iteration (also called a pass) over all objects. Using the new object number i, go to step 2, unless a whole pass has already been made since the last time i was tested, without any reduction in the value of the objective function. In the latter case, stop with the end partition achieved at that point. Before returning to step 2, it is also necessary to test that $t \leqslant$ ITMAX. If not, stop.

For tr W, the exchange condition is easy to test. For a given i with $i \in C_p$ the exchange is successful if and only if there exists at least one $j \neq p$ with

$$\frac{m_j}{m_j + 1} \| x_i - \bar{x}_j \|^2 < \frac{m_p}{m_p - 1} \| x_i - \bar{x}_p \|^2 \tag{2.51}$$

Fortunately, the following is true.

**Theorem 2.7**: For ITMAX = $\infty$, that is, without limiting the number of passes, the exchange method produces a minimal distance partition. If this partition is used as a starting partition for the minimal distance method, it will almost always remain unchanged.

*Proof:* After the exchange method has been applied, (2.51) implies that, for each object $i \in M$ with $i \in C_p$,

$$\frac{m_j}{m_j + 1} \| x_i - \bar{x}_j \|^2 \geqslant \frac{m_p}{m_p - 1} \| x_i - \bar{x}_p \|^2 \quad (j \neq p) \tag{2.52}$$

holds true. (For $m_p = 1$, this is trivially true since the right hand side is then zero.) If (2.52) is multiplied by $(m_j + 1)/m_j > 1$ on the left, and by $(m_p - 1)/m_p < 1$ on the right, the inequality is preserved. Unless $x_i = \bar{x}_p = \bar{x}_j$, the inequality becomes a strict one. In any case, it is true that for each $i \in C_p$ and all $j \neq p$

$$\| x_i - \bar{x}_p \| \leqslant \| x_i - \bar{x}_j \| , \tag{2.53}$$

In other words, the partition possesses the minimal distance property defined in (2.45). Excluding the case $x_i = \bar{x}_p = \bar{x}_j$ which occurs only in artificially constructed examples, we have the strict inequality

$$\| x_i - \bar{x}_p \| < \| x_i - \bar{x}_j \| . \tag{2.54}$$

In this situation, step 3 of the minimal distance makes no further reassignment. (For $m_p = 1$, these conclusions remain valid since the left hand side of (2.53) and (2.54) is then zero.)

The theorem leaves open the reverse possibility of applying the exchange method to an end partition of the minimal distance method and thereby reducing the objective function. Indeed, this is a constructive possibility, as will be seen in sections 11.3 and 11.4.

In a similar way to theorem 2.7, we can prove the following theorem.

**Theorem 2.8**: Using the minimum variance criterion, every optimal partition is also a minimal distance partition. In other words, the minimal distance property is a necessary, though not sufficient, condition for optimality.

*Proof:* There is nothing to prove for $n=1$ or $n=m$. Now let $1 < n < m$ and let

$C = (C_1, \ldots, C_n)$ be optimal. If C were not a minimal distance partition, there would exist an $i \in C_p$ with $m_p > 1$ and a $j \neq p$ with

$$\| x_i - \bar{x}_j \| < \| x_i - \bar{x}_p \| . \tag{2.55}$$

Now square (2.55) on both sides, multiply the left by $m_j/(m_j+1) < 1$ and the right by $m_p/(m_p-1) > 1$. The result is (2.51) which contradicts optimality as it would allow a successful exchange.

Of course, the minimal distance and exchange methods are not the only approximative methods for our problems of combinatorial optimization, of which tr W is a first example. The exchange method, in particular, has many variants.

For example, when calculating the quantities $f_j$ according to (2.49), it is possible to stop at the first j with $f_j > 0$, and to perform the exchange immediately [B40]. Thus, less than maximum effort may be used to achieve a less than maximum reduction in the objective function. It is also possible to add at the beginning of the usual method an attempt to simultaneously exchange two objects from different classes [B4]. The simultaneous exchange of several objects has also been considered [B27]. Further, it is possible in each sequential pass through the objects to skip those whose class membership was unchanged at the previous pass [B25]. Finally, combinations of these and other procedures are conceivable, as well as a change in the number of classes during the iterations.

Neither for tr W nor for the other objective functions will we go into details of these various implementations. There are several reasons for this. Firstly, it is not always obvious or proven that the end partitions produced by the various methods do have the minimal distance property which is a necessary condition for an optimal partition. Secondly, the programming, computer time, and storage requirements for these variants are always higher, though it cannot be proved that they produce a better end partition (with a smaller value of the objective function) than the minimal distance or exchange methods. Finally, it is simple and economical to apply the exchange algorithm several times with different starting partitions and thus to arrive at a distribution of end partitions which will allow their quality to be judged. This will be shown in the examples later on. If an algorithm requiring more computer time is used, less of these valuable repeats are possible.

In addition to the approximative methods described, there exist numerous attempts at creating algorithms which actually produce optimal partitions. They tend to be based on different formulations of the objective function, not only in the case of tr W. Linear integer and dynamic programming [A4, B3, A8], branch and bound methods [B3, B26, B28], and even a non-linear optimization model [A4, B3, B23] have been discussed. However, the indications are that these methods, insofar as they, too, are approximative methods, cannot be regarded as better or more efficient. Insofar as they find an optimal partition, they can only be applied for very small numbers of objects m if the computation time is to be kept within reasonable and predictable bounds. Hence, they are not useful in practice. Since this argument holds – so far as the analogy is possible –

for the other objective functions mentioned in later chapters whose construction is more complicated than tr W, this book will in general be restricted to minimal distance and exchange methods.

## 2.9 Separating hyperplanes

One of the elementary geometric ideas associated with a sensible dissection is that, if a partition is optimal with respect to some objective function, then the convex hulls of different clusters do not intersect, that is, they are disjoint except for points on their boundary. Indeed, we have the following theorem.

**Theorem 2.9:** Let the variance criterion be the objective function. For each pair of classes in a minimal distance partition with distinct mean vectors, and in particular in an optimal partition, there exists a separating hyperplane which contains at most those $x_i$ whose distances to the two means are equal.

*Proof:* The set of all points $x \in \mathbb{R}^s$, which have the same Euclidean distance from $\bar{x}_p \neq \bar{x}_q$, is

$$\{x : \| x - \bar{x}_q \| = \| x - \bar{x}_p \| \} \tag{2.56}$$

Boundary points with $\| x_i - \bar{x}_p \| = \| x_i - \bar{x}_q \|$ and the case of $m_p = m_q = 1$ may be omitted without loss of generality. Hence, we can assume that $m_p > 1$. We assert that (2.56) is the separating hyperplane. If this were not so, there would exist $i, k \in C_p$ with $i \neq k$ such that

$$\| x_i - \bar{x}_q \| < \| x_i - \bar{x}_p \| \tag{2.57}$$

and

$$\| x_k - \bar{x}_q \| > \| x_k - \bar{x}_p \| . \tag{2.58}$$

(i and k might have to be interchanged to make (2.57) and (2.58) true). However, (2.57) contradicts the minimal distance property. It remains to be shown that (2.56) defines a hyperplane. Squaring both sides and substituting

$$\| \bar{x}_q \|^2 - \| \bar{x}_p \|^2 = (\bar{x}_q - \bar{x}_p)^T (\bar{x}_q + \bar{x}_p),$$

we obtain

$$(x - \frac{1}{2}(\bar{x}_q + \bar{x}_p))^T (\bar{x}_q - \bar{x}_p) = 0 \tag{2.59}$$

as required. The plane is perpendicular to the line from $\bar{x}_p$ to $\bar{x}_q$, intersecting it at its mid point.

**Problem 2.1**: For $C_i \subset C_p$ let $C_q := C_p \backslash C_i$, and for $C_i \not\subset C_p$ let $C_q := C_p \cup C_i$ with $m_i > 1$. ($C_i = \{i\}$ and $m_i = 1$ is identical to the case dealt with above). Develop formulae for the mean vectors $\bar{x}_q$, for the scatter matrices $W_q$ and for the quantities $e(C_q)$, expressing them in the simplest possible way as functions of the corresponding quantities for $C_p$ and $C_i$.

**Problem 2.2**: Demonstrate that, for $n = 2$, the minimum variance criterion is equivalent to

$$\max_{C \in P(2,M)} D(C_1, C_2)$$

with

$$D(C_1, C_2) := \frac{m_1 m_2}{m} \| \bar{x}_1 - \bar{x}_2 \|^2 .$$

In addition, prove that, for an optimal partition $(C_1, C_2)$ and $x_i$ with $i \in C_2$ and $m_2 > 1$,

$$\| x_i - (\bar{x} - \frac{m_1 m_2}{m} (\bar{x}_1 - \bar{x}_2)) \|^2 \leqslant \frac{m_1 m_2 (m_1+1)(m_2-1)}{m^2} \| \bar{x}_1 - \bar{x}_2 \|^2 .$$

In other words, prove that all $x_i$ with $i \in C_2$ lie within a sphere. Expressing the objective function in this way, can you think of a new approximative method for finding optimal partitions?

**Problem 2.3**: What is the meaning of the separation theorem 2.9 for $s = 1$? What do the partitions look like if you assume without loss of generality that $x_1 \leqslant x_2 \leqslant \ldots \leqslant x_m$? What are the separating hyperplanes? For this special situation, can you think of a heuristic, or even an exact method for minimizing the variance criterion?

# 3 The minimum determinant criterion

## 3.1 Derivation from an optimality principle for an adaptive metric

In summary, the important attractive properties of the minimum variance criterion are:

- simple calculation of class means using (2.3);
- simple formulae for up- and downdating means and sums of squared deviations, as objects are added to or removed from a class (2.35), (2.39), (2.37) and (2.41);
- Theorem 2.2, stating that minimizing the sum of the variation within the classes is equivalent to maximizing a measure of the heterogeneity between the classes;
- monotonicity according to Theorem 2.5;
- Theorem 2.8 stating that the minimal distance property is a necessary condition for optimality of a partition; and the
- separation property according to Theorem 2.9.

A drawback of the variance criterion is that it is only invariant with respect to orthogonal transformations and translations, but not with respect to simple changes of scale (2.29) which are of practical importance.

In this chapter we shall derive a criterion which is invariant [B18] not only with respect to scale transformations, but also with respect to the much larger class of regular transformations and translations. The price to pay for this is a much more complicate set of up- and downdating formulae to be applied when objects are added to or taken from a class, and there appears to be no analogy to Theorem 2.2. Fortunately, the other positive properties of the variance criterion will be preserved.

We start by considering that the Euclidean metric

$$d(x,y) := \| x - y \| = \sqrt{(x-y)^T(x - y)} \tag{3.1}$$

chosen for the variance criterion is a very special one. It is well known that, for positive definite matrices G,

$$d_G(x,y) := \| x - y \|_G = \sqrt{(x - y)^T G(x - y)} \tag{3.2}$$

also defines a metric. G may be chosen to suit our purpose. For a fixed y, $\{x : \| x-y \| \leqslant c\}$ is a sphere with radius c, and, for $G \neq I$, $\{x : \| x-y \|_G \leqslant c\}$ is an ellipsoid.

We construct sums of squared deviations analogous to (2.2):

$$S_G(y) := \sum_{i \in C_j} \| x_i - y \|_G^2 = \sum_{i \in C_j} (x_i - y)^T G (x_i - y) \quad . \tag{3.3}$$

As previously, there are minimized by the mean vectors $\bar{x}_j$. This follows from

$$\frac{\partial S_G}{\partial y} = -2 \sum_{i \in C_j} G(x_i - y)$$

and from the necessary condition $\frac{\partial S_G}{\partial y} = 0$, since G is regular and since

$$\frac{\partial^2 S_G}{\partial y^2} = 2 m_j G$$

is positive definite.

Now let us consider the transformed variance criterion

$$D_G(C) = \sum_{j=1}^{n} \sum_{i \in C_j} (x_i - \bar{x}_j)^T G (x_i - \bar{x}_j) \tag{3.4}$$

initially for a fixed partition C of length n. We wish to determine G so that expression (3.4) is optimized. This, however, only makes sense if appropriate additional conditions are imposed. For instance, $G = \text{diag}(\epsilon, \epsilon, \ldots, \epsilon)$ would remain positive definite for $\epsilon \to 0$, $\epsilon \neq 0$, but expression (3.4) could be made arbitrarily small. Thus, for a given G, we must exclude all positive definite matrices $G_a := aG$, where $a > 0$ and $a \neq 1$.

An appropriate norm function must take account of all matrix elements, hence tr G is not suitable when $s > 1$. Without doubt, det G is a suitable norm function. We therefore choose as our additional condition that

$$(\det G)^{1/s} = 1 \tag{3.5}$$

In Chapter 5 we shall consider other possibilities. To find an optimum of (3.4) under the auxiliary condition (3.5), we form the Lagrangian function

$$F(G,\lambda) = \sum_{j=1}^{n} \sum_{i \in C_j} (x_i - \bar{x}_j)^T G (x_i - \bar{x}_j) - \lambda\, ((\det G)^{1/s} - 1), \tag{3.6}$$

and calculate its derivatives (see [B5])

$$\frac{\partial F}{\partial G} = \sum_{j=1}^{n} \sum_{i \in C_j} (x_i - \bar{x}_j)(x_i - \bar{x}_j)^T - \frac{1}{s}(\det G)^{1/s} G^{-1} \lambda \tag{3.7}$$

and

$$\frac{\partial F}{\partial \lambda} = -((\det G)^{1/s} - 1) \; . \tag{3.8}$$

We equate the derivatives to zero. Using (2.15) and (2.19), (3.7) leads in turn to

$$W = \frac{1}{s} \lambda G^{-1} (\det G)^{1/s} \, ,$$

$$\frac{1}{s} \lambda I = GW,$$

$$\lambda^s = s^s \det(GW) = s^s \det W,$$

$$\lambda = s(\det W)^{1/s} \, ,$$

and finally to

$$G = (\det W)^{1/s} W^{-1} \, , \tag{3.9}$$

assuming $W^{-1}$ to exist and to be positive definite. If this is the case, then function (3.4) which is linear in G assumes an optimum when G is given by (3.9), since the auxiliary condition (3.5) is a concave function on the set of positive definite matrices [B30]. Provided it exists, the optimum is unique.

## 3.2 The positive definiteness of scatter matrices

Before we formally formulate the new objective function, we will examine the conditions which partitions must fulfil in order to ensure that W is positive definite. Unfortunately we shall find only necessary, and no sufficient conditions which are easily verifiable.

According to (2.19), W is the sum of the scatter matrices $W_j$ defined in (2.15). These are always positive semi-definite, since for $y \in \mathbb{R}^s$, $y \neq 0$,

$$y^T W_j y = y^T \Big( \sum_{i \in C_j} (x_i - \bar{x}_j)(x_i - \bar{x}_j)^T \Big) y = \sum_{i \in C_j} (y^T (x_i - \bar{x}_j))^2 \geqslant 0$$

holds. As a sum of matrices with this property, W is also at least positive semi-definite. W may even be positive definite when all the $W_j$ are positive semi-definite. This is shown by the example

$$W_1 = \begin{pmatrix} 1 & 2 \\ 2 & 4 \end{pmatrix}, \qquad W_2 = \begin{pmatrix} 4 & 2 \\ 2 & 1 \end{pmatrix}.$$

However, W is certainly positive definite if at least one of the $W_j$ is positive definite. Thus, we are asking which conditions the vectors $x_i - \bar{x}_j$ must satisfy to ensure that for some j and $y \neq 0$

$$\sum_{i \in C_j} (y^T(x_i - \bar{x}_j))^2 > 0$$

always holds true. This is equivalent to saying that there is no non-trivial linear combination of the vectors $x_i - \bar{x}_j$ which adds up to the null vector. That in turn is possible exactly when the $x_i - \bar{x}_j$ are linearly independent. The following theorem makes a statement about the range of $m_j$ which does not guarantee this.

**Theorem 3.1**: Let $y_1, \ldots, y_r \in \mathbb{R}^s$ with $1 \leqslant r \leqslant s$ be linearly independent, and let $\bar{y} = \frac{1}{r}(y_1 + \ldots + y_r)$ be the mean vector. Then the vectors $y_i - \bar{y}$ $(i=1, \ldots, r)$ are linearly dependent.

*Proof:* As usual, we assume that there is a linear combination adding up to the null vector, and we show that not all coefficients vanish. Let

$$\sum_{i=1}^{r} \beta_i(y_i - \bar{y}) = 0$$

for $\beta_i \in \mathbb{R}$ $(i=1, \ldots, r)$. Substituting the definition of $\bar{y}$, we get

$$\sum_{i=1}^{r} (r\beta_i - (\beta_1 + \ldots. + \beta_r))y_i = 0.$$

But, since the $y_i$ are linearly independent,

$$r\beta_i - (\beta_1 + \ldots + \beta_r) = 0 \quad (i=1, \ldots, r)$$

must hold. However, the coefficient matrix of this system of linear equations for $\beta_1, \ldots, \beta_r$ is singular since, for example, the sum of the first $(r-1)$ row vectors is equal to minus the $r^{th}$ row vector. Thus, this homogeneous system of linear equations does possess solutions $(\beta_1, \ldots, \beta_r)^T \neq 0$.

Related to the matrix $W_j$, the negation of Theorem 3.1 provides the following corollary.

*Corollary:* If the $m_j$ vectors $x_i - \bar{x}_j$ $(i \in C_j)$ are linearly independent, that is, if $W_j$ is positive definite, then $m_j > s$.

Whether, conversely, $m_j > s$ implies that $W_j$ is positive definite, depends in a complicated way on the vectors $x_i$ and cannot be dealt with simply. Overall, we have the following situation: Using the notation

$$P^+(n,M) = \{C : C \in P(n,M),\ W = W(C) \text{ positive definite}\} \tag{3.10}$$

$$P_1(n,M) = \{C : C \in P(n,M),\ W_j \text{ positive semidefinite but not positive definite for } j = 1, \ldots, n, \text{ with } W \text{ positive definite}\} \tag{3.11}$$

$$P_2(n,M) = \{C : C \in P(n,M),\ W_j \text{ positive definite for at least one } j \in \{1, \ldots, n\}\} \tag{3.12}$$

$$P_3(n,M) = \{C : C \in P(n,M),\ m_j > s \text{ for all } j = 1, \ldots, n\} \tag{3.13}$$

$$P_4(n,M) = \{C : C \in P(n,M),\ m_j > s \text{ for at least one } j \in \{1, \ldots, n\}\}, \tag{3.14}$$

we have

$$P^+(n,M) \subset P_1(n,M) \cup P_2(n,M) \tag{3.15}$$

and

$$P_2(n,M) \subset P_3(n,M) \subset P_4(n,M). \tag{3.16}$$

Whether a partition C belongs to $P^+(n,M)$ can in general only be determined numerically and hence, because of the inevitable rounding errors, not with absolute certainty. In practical applications, $P_1(n,M)$ will frequently be empty. There will be an $n_0 < m$, starting from which $P^+(n_0,M) = \phi$.

### 3.3 Re-formulation of the transformed variance criterion

If we examine expression (3.4) with G accordingly (3.9), i.e.

$$D_W(C) = \sum_{j=1}^{n} \sum_{i \in C_j} (x_i - \bar{x}_j)^T (\det W)^{1/s} W^{-1} (x_i - \bar{x}_j), \tag{3.17}$$

we find that we have a transformed version of the variance criterion. However, it is expressed in a metric which, since W is dependent on C, is similarly dependent on the partition C. For minimizing (3.17), only partitions in $P^+(n,M)$ are feasible. A partition $C_0 \in P^+(n,M)$ is called optimal relative to (3.17) if

$$D_W(C_0) = \min_{C \in P^+(n,M)} D_W(C) \tag{3.18}$$

holds true. In addition, we admit only those n for which $P^+(n,M) \neq \phi$.

Written out in full, the optimization problem

$$\min_{C \in P(n,M)} \ \min_{\substack{G \text{ pos. def.} \\ (\det G)^{1/s}=1}} \ \sum_{j=1}^{n} \ \min_{y_j \in \mathrm{IR}^s} \ (x_i - y_j)^T G (x_i - y_j) \tag{3.19}$$

is now reduced to (3.18). We can simplify $D_W(C)$ even further. For, using G as defined in (3.9), we get

$$\begin{aligned} D_W(C) &= \operatorname{tr} D_W(C) \\ &= \sum_{j=1}^{n} \sum_{i \in C_j} \operatorname{tr}((x_i - \bar{x}_j)^T G(x_i - \bar{x}_j)) \\ &= \sum_{j=1}^{n} \sum_{i \in C_j} \operatorname{tr}((x_i - \bar{x}_j)(x_i - \bar{x}_j)^T G) \qquad (3.20) \\ &= \sum_{j=1}^{n} \operatorname{tr}(W_j G) \\ &= \operatorname{tr}(\sum_{j=1}^{n} W_j G) \\ &= \operatorname{tr}(WG) \\ &= \operatorname{tr}((\det W)^{1/s} I) \\ &= s(\det W)^{1/s} . \end{aligned}$$

Since the optimization of $D_W(C)$ and of det W = det W(C) is thus shown to be equivalent, we can also seek an optimal partition according to

$$\det W(C_0) = \min_{C \in P^+(n,M)} \det W(C) . \qquad (3.21)$$

This explains why (3.21) and thus also (3.18) is called the determinant criterion. What lies behind this name is made explicit by the formulation (3.19). In the same way that we used 'tr W' for the variance criterion, we shall now use 'det W' as abbreviation for the determinant criterion. In both instances, the notation disguises the dependence on C.

### 3.4 Invariance properties

Since the metric underlying the determinant criterion depends, according to (3.9), on the partition, we cannot expect a relationship analogous to (2.17) whose left hand side is independent of the partition. Thus, there is no analogue to Theorem 2.2.

On the other hand, the determinant criterion, unlike the variance criterion, does have the desirable property of invariance relative to scale transformations (2.29) and even relative to all regular transformations and translations

$$x_i' = Ax_i + b \ (i=1, \ldots, m), \ A \in \mathbb{R}(s,s) \text{ non-singular}, \ b \in \mathbb{R}^s . \qquad (3.22)$$

We have the following theorem.

**Theorem 3.2:** For every feasible partition $C \in P^{+}(n,M)$,

$$\det W'(C) = (\det A)^2 \det W(C) \ , \tag{3.23}$$

where $W'$ is formed in the same way as W, but from vectors $x_i'$ which have been transformed according to (3.22).

*Proof:* Because of (2.27) we have

$$x_i' - \bar{x}_j' = A(x_i - \bar{x}_j)$$

and consequently

$$\begin{aligned} W' &= \sum_{j=1}^{n} \sum_{i \in C_j} (x_i' - \bar{x}_j')(x_i' - \bar{x}_j')^T \\ &= \sum_{j=1}^{n} \sum_{i \in C_j} A(x_i - \bar{x}_j)(x_i - \bar{x}_j)^T A^T \\ &= AWA^T \ . \end{aligned} \tag{3.24}$$

The theorem follows from taking determinants of both sides. Since A is regular and W is positive definite, $W'$ is positive definite as well. Thus, for all feasible partitions, det W and det $W'$ differ only by a constant factor which, for regular transformations, is non-zero. Independent of the transformation chosen, there will exist optimal partitions with the same composition.

Regarding the invariance property, one could argue that the step from the variance to the determinant criterion had been too large, and that it might be sensible to look for criteria which are invariant only with respect to scale transformations H as in (2.29) and translations. However, it is shown [B15] that metrics with the property

$$d(Hx + b, Hy + b) = d(x,y)$$

depend only on which components of x and y differ, and not on the size of the difference. Such criteria would therefore involve a great loss of information and are thus of no use in practice. That is also the reason why it is in general unwise to construct objective functions which merely take account of invariance properties.

## 3.5 Up- and downdating formulae for determinants and inverses of sums of dyadic products

An important factor in using the variance criterion is the fact that the mean vectors and sums of squared deviations can be updated easily as an object is exchanged between two classes. In (2.36) and (2.40) we have already derived appropriate formulae for the scatter matrices. Since we are now dealing with the determinant of a sum of scatter matrices, and since expression (3.17) involves the inverse of W, the following proposition will be useful.

**Theorem 3.3**: Let $y \in \mathbb{R}^s$, let $A \in \mathbb{R}(s,s)$ be invertible, and let $\beta \in \mathbb{R}$. Then the following holds (see [B37]):

(a) $\det(A + \beta yy^T) = \det A\,(1 + \beta y^T A^{-1} y)$. (3.25)

(b) The inverse of $A + \beta yy^T$ exists whenever

$$1 + \beta y^T A^{-1} y \neq 0 \tag{3.26}$$

is true. In that case,

$$(A + \beta yy^T)^{-1} = A^{-1} - \frac{\beta A^{-1} yy^T A^{-1}}{1 + \beta y^T A^{-1} y} . \tag{3.27}$$

*Proof:* The identity

$$\begin{pmatrix} 1 & 0 \\ \beta y & A+\beta yy^T \end{pmatrix} \begin{pmatrix} 1 & -y^T \\ 0 & I \end{pmatrix} = \begin{pmatrix} 1 & -y^T \\ 0 & A \end{pmatrix} \begin{pmatrix} 1+\beta y^T A^{-1} y & 0 \\ \beta A^{-1} y & I \end{pmatrix}$$

may be verified by multiplying out both sides. Forming the determinant of both sides results in (a). Statement (b) may be verified by multiplying both sides of (3.27) by $A + \beta yy^T$.

## 3.6 Monotonicity

For the variance criterion, the proof of monotonicity according to Theorem 2.5 was quite simple. It relied on the fact that, for an optimal partition $C \in P(n,M)$, a partition

$$C'' = (C_1, \ldots, C_p \setminus \{i\}, \ldots, C_n, C_{n+1} := \{i\}) \tag{3.28}$$

is a member of $P(n+1,M)$ and has a corresponding value of the objective function which is no greater than that for C. If we want to apply the same procedure using the determinant criterion, we run into difficulties if $C \in P^+(n,M)$ but $C'' \notin P^+(n+1,M)$. Hence we must establish appropriate assumptions.

**Theorem 3.4**: For given vectors $x_i$ $(i=1, \ldots, m)$ and for some n with $1 \leqslant n < m$, let the sets $P^+(n,M)$ and $P^+(n+1,M)$ be non-empty. Assume that for every optimal partition $(C_1, \ldots, C_n) = C \in P^+(n,M)$ there exists an $i \in M$ such that $i \in C_p$ with $m_p > 1$, and such that $C''$, defined by (3.28), is a member of $P^+(n+1, M)$. If $C' \in P^+(n+1, M)$ is a partition that is optimal with respect to the determinant criterion, then the following inequality holds:

$$\det W(C') \leqslant \det W(C) . \tag{3.29}$$

*Proof:* By hypothesis, W(C) and W(C″) are positive definite. Hence,

$$\det W(C) > 0, \quad \det W(C'') > 0 . \tag{3.30}$$

Since $W_{n+1} = 0$, the relationship

$$W(C'') = W(C) - \frac{m_p}{m_p - 1} (x_i - \bar{x}_p)(x_i - \bar{x}_p)^T , \tag{3.31}$$

holds. Using Theorem 3.3, we obtain

$$\det W(C'') = (1-c) \det W(C) \tag{3.32}$$

with

$$c = \frac{m_p}{m_p - 1} (x_i - \bar{x}_p)^T W(C)^{-1} (x_i - \bar{x}_p) \tag{3.33}$$

If $x_i = \bar{x}_p$, then $c = 0$ and we have

$$\det W(C'') = \det W(C). \tag{3.34}$$

(3.30) and (3.32) together imply that $1 - c > 0$. When $x_i \neq \bar{x}_p$, (3.33) and the fact that W(C) is positive definite imply that $c > 0$, and thus that

$$0 < c < 1 \quad (x_i \neq \bar{x}_p) \tag{3.35}$$

In this case,

$$\det W(C'') < \det W(C). \tag{3.36}$$

Since C′ is optimal, (3.34) and (3.36) lead to

$$\det W(C) \geqslant \det W(C'') \geqslant \det W(C') > 0.$$

The significance of the condition $C'' \in P^+(n,M)$ is the following. It must be possible to take a class containing several objects, which forms part of an optimal partition C, and to remove one object into a class of its own, while retaining the positive definiteness of the matrix W(C), which is thereby reduced by one dyadic product. Since W(C) is the sum of m dyadic products, its positive definiteness is not generally affected when $m \gg s$. The assumption is not, therefore, very restrictive. In essence, the situation is similar to that found with the variance criterion. Assuming the necessary conditions are fulfilled, increasing the number of classes will lead to a reduction in the determinant criterion unless the removed object happens to coincide with one of the class means.

### 3.7 Minimal distance method

From the determinant criterion as stated in (3.17) with $C \in P^+(n,M)$, it is clear that we can define a minimal distance partition relative to det W(C), analogous to (2.47), as

$$i \in C_j \Longleftrightarrow \| x_i - \bar{x}_j \|_{W^{-1}} = \min_{k=1,\dots,n} \| x_i - \bar{x}_k \|_{W^{-1}}. \tag{3.37}$$

Here we made use of the fact that, in measuring distances, for a fixed partition C, the factor $(\det W)^{1/s}$ may be omitted from $G = (\det W)^{1/s} W^{-1}$, and $W^{-1}$ may be used in place of G.

The minimal distance method is very similar to that used with the variance criterion, except that the norm used at the $t^{th}$ step is $\| \, . \, \|_{W^{-1}}$ instead of $\| \, . \, \|$, where $W = W(C^{(t)})$. Technically, it is first necessary to perform a Cholesky decomposition [B54]

$$W(C^{(t)}) = LL^T \tag{3.38}$$

using a lower triangular matrix L, and to calculate the squared distances according to

$$(x_i - \bar{x}_j^{(t)})^T \, W(C^{(t)})^{-1} \, (x_i - \bar{x}_j) = \| L^{-1}(x_i - \bar{x}_j^{(t)}) \|^2 \tag{3.39}$$

where the right hand side of (3.39) consists of the ordinary Euclidean distance of the vectors $x_i$ and $\bar{x}_j^{(t)}$ after transformation by $L^{-1}$. Thus, the $t^{th}$ step of the algorithm involves the solution of the mn systems of linear equations

$$Lz_{ij} = x_i - \bar{x}_j^{(t)} \tag{3.40}$$

for $z_{ij} \in \mathbb{R}^s$ $(i=1, \dots, m, j=1, \dots, n)$. Because L is triangular, this is done by simple forward substitution.

In the case of the variance criterion, empty classes could occur in (2.47). The same can be the case in the present context. In addition, it can of course happen that $C^{(t+1)} \notin P^+(n,M)$, which makes it impossible to continue with the iterations. Consequently, the theorem about the contraction property of the minimal distance method holds only with the corresponding restriction stated in the following theorem.

**Theorem 3.5**: If the starting partition is in $P^+(n, M)$ and if the sequence of minimal distance partitions according to (2.47), but using the $W(C^{(t)})^{-1}$ metric, is also in $P^+(n,M)$, then the minimal distance algorithm for det W will also lead to a monotonic reduction in the values of the objective function.

*Proof:* With $G^{(t)} = (\det W(C^{(t)}))^{1/s} \, W(C^{(t)})^{-1}$ we have

$$s(\det W(C^{(t)}))^{1/s} = \sum_{j=1}^{n} \sum_{i \in C_j^{(t)}} (x_i - \bar{x}_j^{(t)})^T G^{(t)} (x_i - \bar{x}_j^{(t)}) \tag{3.41}$$

$$\geqslant \sum_{j=1}^{n} \sum_{i \in C_j^{(t)}} \min_{k=1,\ldots,n} (x_i - \bar{x}_k^{(t)})^T G^{(t)} (x_i - \bar{x}_k^{(t)})$$

$$= \sum_{j=1}^{n} \sum_{i \in C_j^{(t+1)}} (x_i - \bar{x}_j^{(t)})^T G^{(t)} (x_i - \bar{x}_j^{(t)})$$

$$\geqslant \sum_{j=1}^{n} \sum_{i \in C_j^{(t+1)}} (x_i - \bar{x}_j^{(t+1)})^T G^{(t)} (x_i - \bar{x}_j^{(t+1)})$$

$$\geqslant \sum_{j=1}^{n} \sum_{i \in C_j^{(t+1)}} (x_i - \bar{x}_j^{(t+1)})^T G^{(t+1)} (x_i - \bar{x}_j^{(t+1)})$$

$$= s(\det W(C^{(t+1)}))^{1/s}$$

and hence

$$\det W(C^{(t+1)}) \leqslant \det W(C^{(t)}).$$

## 3.8 The exchange method and a necessary condition for optimal partitions; separating hyperplanes

Basically, the exchange method is the same as that for the variance criterion. Starting from a feasible initial partition $(C_1, \ldots, C_n) = C \in P^+(n,M)$ all objects are scanned in sequence, and for each $i \in C_p$ with $m_p > 1$, all $n-1$ partitions

$$C'' = (C_1, \ldots, C_p \setminus \{i\}, \ldots, C_j \cup \{i\}, \ldots, C_n) \quad (j \neq p) \tag{3.42}$$

are examined for

$$C'' \in P^+(n,M) \tag{3.43}$$

and for

$$\det W(C'') < \det W(C). \tag{3.44}$$

Compared to the variance criterion, the additional test for (3.43) or, equivalently,

$$\det W(C'') > 0 \tag{3.45}$$

needs to be performed. A relatively simple condition for making an exchange, analogous to (2.51), is given by Theorem 3.3. Using the notation

$$C' = (C_1, \ldots, C_p \setminus \{i\}, \ldots, C_n) \tag{3.46}$$

the following two equations follow from (2.36) and (2.40):

$$W(C) = W(C') + \frac{m_p}{m_p - 1} (x_i - \bar{x}_p)(x_i - \bar{x}_p)^T, \tag{3.47}$$

$$W(C'') = W(C') + \frac{m_j}{m_j + 1} (x_i - \bar{x}_j)(x_i - \bar{x}_j)^T. \tag{3.48}$$

Condition (3.44) provides the inequality

$$\frac{m_j}{m_j + 1} (x_i - \bar{x}_j)^T W(C')^{-1} (x_i - \bar{x}_j) < \frac{m_p}{m_p - 1} (x_i - \bar{x}_p)^T W(C')^{-1} (x_i - \bar{x}_p), \tag{3.49}$$

which must be satisfied if a successful exchange is to take place. Where (3.49) is satisfied, it is also necessary to check (3.45) to ensure that condition (3.43) is fulfilled.

The exchange condition (3.49) may be rewritten in terms of W(C) rather than W(C′). From (3.47) we have

$$W(C') = W(C) - \frac{m_p}{m_p - 1} (x_i - \bar{x}_p)(x_i - \bar{x}_p)^T.$$

According to Theorem 3.3 this allows $W(C')^{-1}$ to be calculated as a function of $W(C)^{-1}$, provided that the c of (3.33) is non-zero. This, however, is the case because of condition (3.45). After some algebra, and using the abbreviation

$$y_{ki} := x_i - \bar{x}_k\,, \quad k=j,p \tag{3.50}$$

(3.49) is transformed into the inequality

$$\frac{m_j}{m_j + 1} y_{ji}^T W(C)^{-1} y_{ji} < \frac{m_p}{m_p - 1} y_{pi}^T W(C)^{-1} y_{pi} + \frac{m_p}{m_p - 1} \frac{m_j}{m_j + 1} (y_{ji}^T W(C)^{-1} y_{ji} y_{pi}^T W(C)^{-1} y_{pi} - (y_{ji}^T W(C)^{-1} y_{pi})^2). \tag{3.51}$$

The second term on the right hand side is non-negative owing to the generalized Cauchy-Schwarz inequality [A4]. Condition (3.51) can be rewritten in a form which might be more memorable:

$$\left(1 - \frac{m_p}{m_p - 1} y_{pi}^T W(C)^{-1} y_{pi}\right)\left(1 + \frac{m_j}{m_j + 1} y_{ji}^T W(C)^{-1} y_{ji}\right) \tag{3.52}$$

$$+ \frac{m_p}{m_p - 1} \frac{m_j}{m_j + 1} (y_{ji}^T W(C)^{-1} y_{pi})^2 < 1 .$$

We are now ready to demonstrate theorems analogous to 2.7 and 2.8.

**Theorem 3.6**: With an unlimited number of passes, a feasible starting partition, and on the assumption that (3.45) is satisfied for each successful exchange (3.42), the exchange method for the determinant criterion will produce an end partition $C \in P^+(n,M)$ which is a minimal distance partition in the corresponding $W(C)^{-1}$ metric. The associated value of the objective function cannot be further reduced by a step of the minimal distance method.

*Proof:* After the end of the exchange method the opposite of (3.51) holds for every $i \in M$ with $i \in C_p$, $m_p > 1$ and for all $j \neq p$ with $C'' \in P^+(n,M)$; that is,

$$\frac{m_j}{m_j + 1} y_{ji}^T W(C)^{-1} y_{ji}$$

$$\geqslant \frac{m_p}{m_p - 1} y_{pi}^T W(C)^{-1} y_{pi}$$

$$+ \frac{m_p}{m_p - 1} \frac{m_j}{m_j + 1} (y_{ji}^T W(C)^{-1} y_{ji} y_{pi}^T W(C)^{-1} y_{pi} - (y_{ji}^T W(C)^{-1} y_{pi})^2).$$

Considering that the second term on the right hand side is non-negative, we can strengthen the inequality by omitting it. Using (3.50) we obtain

$$\frac{m_j}{m_j + 1} (x_i - \bar{x}_j)^T W(C)^{-1} (x_i - \bar{x}_j)$$

$$\geqslant \frac{m_p}{m_p - 1} (x_i - \bar{x}_p)^T W(C)^{-1} (x_i - \bar{x}_p).$$

Now multiply the left hand side by $(m_j + 1)/m_j > 1$, and the right hand side by $(m_p - 1)/m_p < 1$ and reverse the inequality. The result is

$$\| x_i - \bar{x}_p \|_{W(C)^{-1}} \leqslant \| x_i - \bar{x}_j \|_{W(C)^{-1}} , \quad j \neq p . \tag{3.53}$$

This is the minimal distance property. Except when $x_i = \bar{x}_p = \bar{x}_j$, (3.53) is a strict inequality. The following theorem also holds.

**Theorem 3.7:** Every optimal partition $C \in P^+(n,M)$ under the determinant criterion is a minimal distance partition relative to the $W(C)^{-1}$ metric.

*Proof:* Assume, on the contrary, that C is not a minimal distance partition. Then there exists an $i \in C_p$ with $m_p > 1$ and a $j \neq p$ with

$$\| x_i - \bar{x}_j \|_{W(C)^{-1}} < \| x_i - \bar{x}_p \|_{W(C)^{-1}} . \tag{3.54}$$

Now square both sides of (3.54), multiply the left side by $m_j/(m_j + 1) < 1$, the right by $m_p/(m_p - 1) > 1$, and add on the right

$$\frac{m_p}{m_p - 1} \frac{m_j}{m_j + 1} (y_{ji}^T W(C)^{-1} y_{ji} y_{pi}^T W(C)^{-1} y_{pi} - (y_{ji}^T W(C)^{-1} y_{pi})^2) \geqslant 0.$$

The result is (3.51), implying that an exchange can be made which reduces the objective function. This contradicts the optimality of C, thus proving the theorem.

Finally, we have the following theorem.

**Theorem 3.8:** All minimal distance partitions C under the determinant criterion, and hence all optimal partitions, have the following property. Any two classes in C with different mean vectors can be separated by a hyperplane containing at most those $x_i$ which are equidistant to both mean vectors.

*Proof:* The set of all points with the same $W(C)^{-1}$-distance to $\bar{x}_p$ and $\bar{x}_q$ is given by

$$\{ x : \| x - \bar{x}_q \|_{W(C)^{-1}} = \| x - \bar{x}_p \|_{W(C)^{-1}} \} \tag{3.55}$$

By squaring, we obtain the equation of the hyperplane

$$(x - \frac{1}{2} (\bar{x}_q + \bar{x}_p))^T W(C)^{-1} (\bar{x}_q - \bar{x}_p) = 0. \tag{3.56}$$

The rest of the proof is exactly as for Theorem 2.9, but using the transformed $W(C)^{-1}$ distance for the minimal distance partition $C \in P^+(n,M)$, instead of the Euclidean distance.

The majority of the properties of the variance criterion are transferable to the determinant criterion. They simply apply under a different metric. However, there is no analogue to Theorem 2.2. On the other hand, the determinant criterion is invariant with respect to all regular transformations and translations, including changes of scale. The minimal distance and exchange methods are computationally more expensive for the determinant criterion than for the variance criterion. The minimal distance property of the variance criterion means that the clusters found or imposed by the minimal distance or exchange methods are spherical. Using the determinant criterion we are producing ellipsoids which can be separated by hyperplanes and which have equal axis orientations that are determined by the eigenvectors of $G = (\det W)^{1/s} W^{-1}$ for end partitions C. In the next chapter we shall deal with an objective function which allows ellipsoids with different axis orientations. However, we shall lose the separation property which seems very worthwhile for the purpose of cluster dissection.

**Problem 3.1:** Can you determine the number of partitions from problem 2.3, and of $P_3(n,M)$ and $P_4(n,M)$ according to (3.13) and (3.14)?

**Problem 3.2:** When T is positive definite, it is known that there exists a regular matrix V such that $VTV^T = I$. Using this fact, show that for n=2, the objective functions det W and tr $VWV^T$ are equivalent. (This means that for n=2 it is possible to use the variance criterion for the vectors $y_i := Vx_i$, $i=1, \ldots, m$, instead of the determinant criterion.)

**Problem 3.3:** Let $w_k$ $(k=1, \ldots, s)$ be the eigenvalues of W. Because of the known properties of the trace and determinant functions we have

$$\operatorname{tr} W = \sum_{k=1}^{s} w_k\,, \qquad \det W = \prod_{k=1}^{s} w_k\,.$$

Do these formulations suggest to you ideas for further objective functions? Can det W be used as an objective if W is allowed to be positive semi-definite?

# 4 The criterion of adaptive distances

## 4.1 Derivation from an optimality principle for metrics adapted to individual classes

Take a partition of objects which are characterized by the vectors $x_i \in \mathbb{R}^s$ ($i=1, \ldots, m$). If we permit class specific metrics

$$d_{G_j}(x,y) = \| x - y \|_{G_j} = \sqrt{(x-y)^T G_j (x-y)} \quad (j=1, \ldots, n) \tag{4.1}$$

with (different) positive definite matrices $G_j$, then the sum of squared deviations in a class as in (3.3), that is,

$$S_{G_j}(y) = \sum_{i \in C_j} \| x_i - y \|_{G_j}, \tag{4.2}$$

is still minimized by the mean vector $\bar{x}_j$.

Our procedure is analogous to that employed with the determinant criterion, but using matrices which are cluster-specific instead of applying to the whole set of objects. For a given partition C we seek, amongst all positive definite matrices $G_j$ satisfying the normalization condition

$$(\det G_j)^{1/s} = 1 \quad (j=1, \ldots, n) \tag{4.3}$$

those for which

$$D_{G_1, \ldots, G_n}(C) = \sum_{j=1}^{n} \sum_{i \in C_j} (x_i - \bar{x}_j)^T G_j (x_i - \bar{x}_j) \tag{4.4}$$

is optimized

The Lagrangian function in this case is

$$F(G_1, \ldots, G_n, \lambda_1, \ldots, \lambda_n) \tag{4.5}$$

$$= \sum_{j=1}^{n} \sum_{i \in C_j} (x_i - \bar{x}_j)^T G_j (x_i - \bar{x}_j) - \sum_{j=1}^{n} \lambda_j ((\det G_j)^{1/s} - 1)$$

$$= \sum_{j=1}^{n} \left( \sum_{i \in C_j} (x_i - \bar{x}_j)^T G_j (x_i - \bar{x}_j) - \lambda_j ((\det G_j)^{1/s} - 1) \right).$$

Since the pairs $(G_j, \lambda_j)$ only occur in the $j^{th}$ term of the summation, the necessary conditions

$$\frac{\partial F}{\partial G_j} = 0, \quad \frac{\partial F}{\partial \lambda_j} = 0 \quad (j=1, \ldots, n)$$

imply, as in (3.7), (3.8) and (3.9), that

$$G_j = (\det W_j)^{1/s} W_j^{-1}, \tag{4.6}$$

where $W_j$ is the scatter matrix (2.15) of the $j^{th}$ cluster.

The existence of the minimal solution (4.6) is assured, provided the inverses of $W_j$ exist for all $j=1, \ldots, n$. As we saw in section 3.2, a necessary (and unfortunately not sufficient) condition for this is $m_j > s$. Hence, we introduce the notation for the set of feasible partitions

$$P^{++}(n,M) = \{C : C \in P(n,M),\ W_j \text{ positive definite for } j=1, \ldots, n\} \tag{4.7}$$

We have $P^{++}(n,M) \subset P^{+}(n,M)$, that is, the set of feasible partitions is further restricted.

An optimal partition $C_o$ relative to (4.4) is then defined by

$$D_{W_1,\ldots,W_n}(C_o) = \min_{C \in P^{++}(n,M)} D_{W_1,\ldots,W_n}(C) \tag{4.8}$$

with

$$D_{W_1,\ldots,W_n}(C) = \sum_{j=1}^{n} \sum_{i \in C_j} (x_i - \bar{x}_j)^T (\det W_j)^{1/s} W_j^{-1} (x_i - \bar{x}_j). \tag{4.9}$$

If we omit the constant factor s, we can write this in the form

$$D_{W_1,\ldots,W_n}(C) = \sum_{j=1}^{n} (\det W_j)^{1/s}. \tag{4.10}$$

as we did in (3.20). The homogeneity in the classes is measured by $(\det W_j)^{1/s}$.

In the search for optimal partitions, the cluster metrics can simultaneously adapt to the position of the points $x_i$, accordingly to (4.9). Hence, (4.9) and (4.10) are referred to as the criterion of adaptive distances [see B12, B13, B14, B31, B35, B46]. For $n=1$, (4.10) is identical to the determinant criterion, except for the exponent $1/s$.

The present optimization problem can be restated analogously to (3.19) as

$$\min_{C \in P(n,M)} \sum_{j=1}^{n} \min_{\substack{G_j \text{ pos. def.} \\ (\det G_j)^{1/s}=1}} \sum_{i \in C_j} \min_{y_j \in \mathbb{R}^s} (x_i - y_j)^T G_j (x_i - y_j). \tag{4.11}$$

We now need to investigate the properties of this objective function.

## 4.2 Properties

The following properties carry over from the determinant criterion to the criterion of adaptive distances. The simple calculation of class centres as mean vectors, the relatively simple formulas for updating the class specific determinants of the scatter matrices, and the invariance with respect to regular transformations and translations, since the $W_j$ are transformed just like W.

The monotonicity property corresponding to Theorem 3.4 is preserved as well. To prove it, one needs to restrict $C''$ to $P^{++}(n,M)$ instead of $P^{+}(n,M)$, take the additivity of (4.10) into account, and carry the exponent $1/s$ along in the proof (see [B31]).

However, there is a price to be paid for the greater flexibility of the class specific metrics which allow ellipsoids with different axis orientations to be found in the course of cluster analysis. No separation theorem analogous to Theorem 2.9 (for the variance criterion) or Theorem 3.8 (for the determinant criterion) continues to be valid, and this is a sacrifice which, in the context of cluster dissection, one would prefer to avoid.

Consider the set of all points which are equidistant from two class mean vectors $\bar{x}_q$ and $\bar{x}_p$ in the respective class specific metrics defined by $G_q = (\det W_q)^{1/s} W_q^{-1}$ and $G_p = (\det W_p)^{1/s} W_p^{-1}$. This is

$$\{x : (x - \bar{x}_q)^T G_q (x - \bar{x}_q) = (x - \bar{x}_p)^T G_p (x - \bar{x}_p)\}. \tag{4.12}$$

For the normal case where $G_q \neq G_p$ this leads not to the equation of a hyperplane but to that of a quadric, that is, of a second order hypersurface. The equation is

$$x^T(G_q - G_p)\,x - 2x^T(G_q\bar{x}_q - G_p\bar{x}_p) + \bar{x}_q^T G_q \bar{x}_q - \bar{x}_p^T G_p \bar{x}_p = 0. \tag{4.13}$$

When $G_q - G_p$ is positive definite, the separating surface has the character of an ellipse, otherwise that of a hyperbola. An example will illustrate the meaning of this.

Let $\bar{x}_1 = (0, 0)^T$ and $\bar{x}_2 = (2, 0)^T$. Let the metrics associated with these two cluster mean vectors be given by the matrices

$$G_1 = \begin{pmatrix} 1/25 & 0 \\ 0 & 1/5 \end{pmatrix}, \qquad G_2 = \begin{pmatrix} 1/5 & 0 \\ 0 & 1/25 \end{pmatrix}.$$

we have $\det G_1 = \det G_2$. (Note that the choice of one as the norm in (4.3) is arbitrary.) Using the notation $x = (a, b)^T$, (4.13) becomes the equation of the hyperbola

$$b^2 = (a - \frac{5}{2})^2 - \frac{5}{4},$$

which is drawn in Fig. 4.1 together with the two mean vectors.

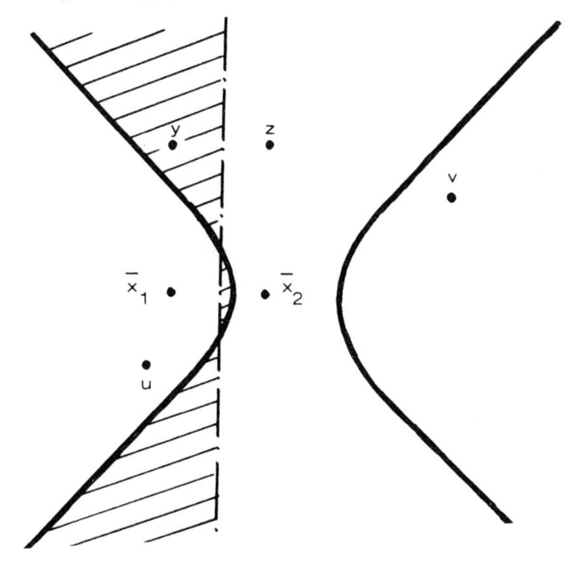


Fig. 4.1

The set of all points in the (a, b) plane of $\mathbb{R}^2$ whose distance from $\bar{x}_1$ in the $G_1$-metric is less than their distance from $\bar{x}_2$ in the $G_2$-metric, lies to the left and to the right of the two branches of the hyperbola. The dashed line corresponds to the separating hyperplane which applies under the variance criterion. For points in the shaded area, their Euclidean distance from $\bar{x}_1$ is smaller, or greater, than that from $\bar{x}_2$, while their $G_1$-distance from $\bar{x}_1$ is greater, or smaller, than their $G_2$-distance from $\bar{x}_2$. As can be checked easily, $y = (0, 3)^T$ is, for example, such a point where normal geometric intuition is no longer correct, whereas it does hold in the case of $z = (2, 3)^T$. Now consider the points u and v in the figure, both of which have a smaller $G_1$-distance from $\bar{x}_1$ than their $G_2$-distance from $\bar{x}_2$, and would therefore, in an optimal partition, lie in the region belonging to $C_1$. The straight line joining u and v does not lie wholly within the region belonging to $C_1$. It follows that the convex hulls of cluster member are not necessarily disjoint.

### 4.3 Minimal distance method

Using the criterion of adaptive distances, a minimal distance partition is defined according to

$$i \in C_j \Leftrightarrow \| x_i - \bar{x}_j \|_{G_j} = \min_{k=1,\ldots,n} \| x_i - \bar{x}_k \|_{G_k}. \tag{4.14}$$

As the example of the previous section showed, the normal geometric intuition based on Euclidean distances does not, however, apply, since different metrics are used on the left and right hand sides of (4.14). As before, it is possible to demonstrate (see [B31]) the following theorem.

**Theorem 4.1**: If the starting partition is from $P^{++}(n,M)$, and if the sequence of minimal distance partitions according to (2.47), but using the metrics in (4.14), is also in $P^{++}(n,M)$, then the minimal distance method for the criterion of adaptive distances does lead to a monotonic decrease in the values of the objective function.

*Proof:* Using the notation $G_j^{(t)} = (\det W_j^{(t)})^{1/s} W_j^{(t)-1}$, we have

$$
\begin{aligned}
s \sum_{j=1}^{n} (\det W_j^{(t)})^{1/s} &= \sum_{j=1}^{n} \sum_{i \in C_j^{(t)}} (x_i - \bar{x}_j^{(t)})^T G_j^{(t)} (x_i - \bar{x}_j^{(t)}) \\
&\geqslant \sum_{j=1}^{n} \sum_{i \in C_j^{(t)}} \min_{k=1,\ldots,n} (x_i - \bar{x}_k^{(t)})^T G_k^{(t)} (x_i - \bar{x}_k^{(t)}) \\
&= \sum_{j=1}^{n} \sum_{i \in C_j^{(t+1)}} (x_i - \bar{x}_j^{(t)})^T G_j^{(t)} (x_i - \bar{x}_j^{(t)}) \\
&\geqslant \sum_{j=1}^{n} \sum_{i \in C_j^{(t+1)}} (x_i - \bar{x}_j^{(t+1)})^T G_j^{(t)} (x_i - \bar{x}_j^{(t+1)}) \\
&\geqslant \sum_{j=1}^{n} \sum_{i \in C_j^{(t+1)}} (x_i - \bar{x}_j^{(t+1)})^T G_j^{(t+1)} (x_i - \bar{x}_j^{(t+1)}) \\
&= s \sum_{j=1}^{n} (\det W_j^{(t+1)})^{1/s} .
\end{aligned}
$$

Since $P^{++}(n,M) \subset P^{+}(n,M)$, the minimal distance method will break down even more frequently here than was the case under the determinant criterion, whenever the method generates infeasible partitions (see [B35]).

## 4.4 Generalization of the objective function, exchange method, and further properties

Looking ahead to Chapter 5, let us consider the exchange method for the criterion of adaptive distances, using a somewhat generalized objective function:

$$
D^{(\beta)}_{W_1,\ldots,W_n}(C) = \sum_{j=1}^{n} (\det W_j)^{\beta}, \quad \beta > 0. \tag{4.15}
$$

For $\beta = 1/s$ we have the previous objective function (4.10).

As in (3.20), it is clear that the right hand side of (4.15) is equal to

$$
\frac{1}{s} \sum_{j=1}^{n} \sum_{i \in C_j} (x_i - \bar{x}_j)^T G_j(\beta) (x_i - \bar{x}_j) \tag{4.16}
$$

where

$$G_j(\beta) = (\det W_j)^\beta W_j^{-1} \qquad (j=1, \ldots, n). \tag{4.17}$$

Together with (4.15) this implies that, for $\beta = 0$,

$$\frac{1}{s} \sum_{j=1}^{n} \sum_{i \in C_j} (x_i - \bar{x}_j)^T W_j^{-1} (x_i - \bar{x}_j) = n. \tag{4.18}$$

Since the objective function for $\beta = 0$ is a constant independent of the partition, values of $\beta$ with $\beta \to 0$ are certainly not suitable in (4.15). Moreover, the matrices (4.17) are not derived from an optimality principle unless $\beta = 1/s$. It would therefore seem impossible to derive the contraction property of the minimal distance method analogous to the proof of Theorem 4.1, unless $\beta = 1/s$.

The exchange method, however, can be formulated very simply even for the generalized expression (4.15). In turn, it is necessary to test for all objects $i \in C_p$ whether the exchange condition

$$(\det W'_p)^\beta + (\det W'_j)^\beta < (\det W_p)^\beta + (\det W_j)^\beta \tag{4.19}$$

holds, where $C'_p := C_p \setminus \{i\}$, $C'_j := C_j \cup \{i\}$, and $W'_p$ and $W'_j$ are the corresponding scatter matrices. Where several $j \neq p$ satisfy the condition, the one to be chosen is that which causes the greatest reduction in the objective function (4.15). However, for each p and j it is also necessary to ensure that the partitions formed lie in $P^{++}(n,M)$. In particular, it must be known or confirmed that

$$m_p > s + 1, \tag{4.20}$$

$$\det W'_p > 0, \qquad \det W'_j > 0. \tag{4.21}$$

Since we have

$$W'_p = W_p - \frac{m_p}{m_p - 1} (x_i - \bar{x}_p)(x_i - \bar{x}_p)^T, \tag{4.22}$$

and

$$W'_j = W_j + \frac{m_j}{m_j + 1} (x_i - \bar{x}_j)(x_i - \bar{x}_j)^T, \tag{4.23}$$

it is possible to use Theorem 3.3. With the abbreviations

$$c_p = \frac{m_p}{m_p - 1} (x_i - \bar{x}_p)^T W_p^{-1} (x_i - \bar{x}_p) \tag{4.24}$$

$$c_j = \frac{m_j}{m_j + 1} (x_i - \bar{x}_j)^T W_j^{-1} (x_i - \bar{x}_j) \tag{4.25}$$

we can rewrite (4.19) in the following form:

$$((1 + c_j)^\beta - 1)(\det W_j)^\beta < (1 - (1 - c_p)^\beta)(\det W_p)^\beta. \tag{4.26}$$

In this,

$$c_k \geqslant 0 \;\text{ and }\; c_k = 0 \Longleftrightarrow x_i = \bar{x}_k \quad (k = p, j). \tag{4.27}$$

However,

$$c_p = c_j = 0 \tag{4.28}$$

is not possible, for (4.26) would not permit it. Since $W'_p$ is to be positive definite, the same line of reasoning as in (3.32) to (3.35) requires that

$$c_p < 1 \tag{4.29}$$

must continue to be true.

We shall not use expression (4.26) in the implementation of the exchange method, but it is very useful in proving various theorems. Since the proofs require us to estimate the size of expressions of the form $(1 + c)^\beta$ in (4.26), we shall quote a well known lemma from the field of analysis, called Bernoulli's inequality.

**Lemma**: For all $c \in \mathbb{R}$ with $c > -1$, $c \neq 0$, and for $\beta \in \mathbb{R}$ we have

$$(1 + c)^\beta > 1 + \beta c \text{ when } \beta < 0 \text{ or } \beta > 1, \tag{4.30}$$

$$(1 + c)^\beta < 1 + \beta c \text{ when } 0 < \beta < 1 . \tag{4.31}$$

For $c = 0$ as well as for $\beta = 1$ and arbitrary c, there is equality in both cases.

Now we can proceed to prove, as before, the following theorem.

**Theorem 4.2:** Let the exchange method for (4.15), with $0 < \beta \leqslant 1$, be carried out for an unlimited number of passes and assume that the starting and end partitions lie in $P^{++}(n,M)$. Then the end partition is a minimal distance partition in the sense of (4.14), but with $G_j(\beta)$ as defined in (4.17) replacing $G_j$ $(j = 1, \ldots, n)$. This end partition cannot be further improved by applying a minimal distance method, suitably modified when $\beta \neq 1/s$ (and not necessarily contracting in that case).

*Proof:* After the end of the exchange method the opposite of (4.26) is true for all $i \in C_p$ $(m_p > s)$ and $j \neq p$, that is,

$$((1 + c_j)^\beta - 1)(\det W_j)^\beta \geqslant (1 - (1 - c_p)^\beta)(\det W_p)^\beta. \tag{4.32}$$

Using (4.31), for $0 < \beta < 1$ and $c_j > 0$, the left hand side can be strictly increased, and the right hand side can be strictly decreased since $0 < c_p < 1$, in consequence of (4.27) and (4.29). After division by $\beta > 0$ we obtain

$$c_j (\det W_j)^\beta > c_p (\det W_p)^\beta, \qquad (0 < \beta < 1). \tag{4.33}$$

This holds also when either $c_j$ or $c_p$ is different from 0. The case $c_p = c_j = 0$ cannot arise, according to (4.28). For $\beta = 1$, and hence for the whole range $0 < \beta \leqslant 1$, we still have

$$c_j (\det W_j)^\beta \geqslant c_p (\det W_p)^\beta, \qquad (0 < \beta \leqslant 1). \tag{4.34}$$

Now substitute in (4.34) the full expressions for $c_j$ and $c_p$ given in (4.24) and (4.25), use notation (4.17), and multiply the left hand side by $(m_j + 1)/m_j > 1$ and the right hand side by $m_p/(m_p - 1) < 1$. Swap left and right hand sides of the inequality and we obtain

$$\| x_i - \bar{x}_p \|^2_{G_p(\beta)} < \| x_i - \bar{x}_j \|^2_{G_j(\beta)} \tag{4.35}$$

for all $j \neq p$ and for all i. In other words, we have a minimal distance partition which would not be changed by step 3 of the minimal distance method.

We also have the following theorem.

**Theorem 4.3**: Every optimal partition $C \in P^{++}(n,M)$ is a minimal distance partition, both under the ordinary ($\beta = 1/s$) and under the modified ($0 < \beta \leqslant 1$) criterion of adaptive distances.

*Proof:* If an optimal partition were not a minimal distance partition, then there would exist an $i \in C_p$ $(m_p > s)$ and $j \neq p$, with

$$\| x_i - \bar{x}_j \|_{G_j(\beta)} < \| x_i - \bar{x}_p \|_{G_p(\beta)} \cdot \tag{4.36}$$

Squaring, multiplying the left hand side by $m_j/(m_j + 1) < 1$ and the right by $m_p/(m_p - 1) > 1$, we obtain

$$c_j (\det W_j)^\beta < c_p (\det W_p)^\beta . \tag{4.37}$$

Bounding the inequality by using on the left $c_j > ((1 + c_j)^\beta - 1)/\beta$ and on the right $c_p < (1 - (1 - c_p)^\beta)/\beta$, we obtain the condition for a successful exchange according to (4.26). This contradicts the optimality of the partition in questions, concluding the proof.

Compared with the determinant criterion, the criterion of adaptive distances has, even in its modified form with $0 < \beta \leqslant 1$, lost the property of the clusters in optimal partitions being seperable by hyperplanes. This can be a disadvantage when one is trying to dissect a set of objects, that is, divide it into classes. However, it is an advantage in the search for a structure, assuming this exists and that the number of classes is specified correctly.

**Problem 4.1**: What do the objective functions tr W, det W, and $\sum_{j=1}^{n} (\det W_j)^{1/s}$ look like for $s = 1$?

**Problem 4.2:** Let $x_i \geqslant 0$ $(i=1, \ldots, m)$ and let at least two of the $x_i$ differ from each other. Further, let $\beta > 0$ and

$$M_\beta := \left(\frac{1}{m} \sum_{i=1}^{m} x_i^\beta\right)^{1/\beta}.$$

Specify a function $f_\beta$ which has no zero and for which $M_\beta$ is a unique minimum. (Hint: for $\beta = 1$, $f_1(x) = \sum_{i=1}^{m} (x_i - x)^2$ is such a function.) What do $M_\beta$ and $f_\beta$ look like as $\beta \to 0$?

**Problem 4.3:** In the text, $\beta > 0$ was assumed. Is

$$\max_{C \in P^{++}(n,M)} \sum_{j=1}^{n} (\det W_j)^\beta$$

a useable objective function when $\beta < 0$? What would the exchange method look like? Would Theorems 4.2 and 4.3 hold?

**Problem 4.4:** How does the objective function (4.10) change if instead of the side conditions (4.3)

$$(\det G_j)^{1/s} = h_j \quad \text{with} \quad h_j > 0 \; (j=1, \ldots, n)$$

were required? How would you choose the $h_j$?

# 5 Further criteria for use with quantitative data†

## 5.1 Other adaptive metrics

The minimum determinant criterion and the criterion for adaptive distances arose from the search for one or several optimal metrics for a fixed partition, given certain normalization conditions. It is, of course, possible to use normalizing functions $f : \mathrm{IR}(s, s) \to \mathrm{IR}$ other than the determinant function. They merely need to have the property that they are functions of all the matrix elements of G and that they take on positive values for positive definite matrices G. Since the eigenvalues of G are positive, this property is shared by the determinant which is the product of the eigenvalues as well as by all of its elementary symmetric functions except by $\operatorname{tr} G = \sum_{k=1}^{s} g_{kk}$ for $s > 1$, provided that the sign is suitably modified. Unfortunately, computation of these functions would require explicit knowledge of the eigenvalues or of the coefficients of the characteristic polynomials, since these elementary symmetric functions cannot be expressed as simple functions of the matrix elements, except in the case of det G and tr G.

A different function possessing the required properties is the Euclidean norm

$$\| G \| := \sqrt{\sum_{i=1}^{s} \sum_{k=1}^{s} g_{ik}^2}\,, \tag{5.1}$$

which is very easy to calculate; even easier than the determinant. If one then wishes to determine amongst all positive definite matrices G those which, for a fixed partition C, satisfy

$$\sum_{j=1}^{n} \sum_{i \in C_j} (x_i - \bar{x}_j)^T G (x_i - \bar{x}_j) \quad \to \text{optimum} \tag{5.2}$$

as before, under the subsidiary condition

$$\| G \| = 1 \tag{5.3}$$

†On first reading, this chapter may be omitted.

then one needs to look at the Lagrangian function

$$F(G, \lambda) = \sum_{j=1}^{n} \sum_{i \in C_j} (x_i - \bar{x}_j)^T G (x_i - \bar{x}_j) - \lambda (\| G \| - 1) . \tag{5.4}$$

The condition $\frac{\partial F}{\partial G} = 0$ gives

$$W = \lambda \frac{1}{\| G \|} G. \tag{5.5}$$

Taking the norm on both sides and using subsidiary condition (5.3) results in

$$\lambda = \| W \| \tag{5.6}$$

and finally in

$$G = \frac{1}{\| W \|} W. \tag{5.7}$$

Since $\| \, . \, \|$ is a norm, (5.7) is the unique solution of the minimization problem if W is positive definite, as it was for the determinant criterion.

After some transformations analogous to (3.20), (5.2) becomes

$$\begin{aligned} \operatorname{tr}(WG) &= \operatorname{tr}\left( \frac{1}{\| W \|} W^2 \right) \\ &= \frac{1}{\| W \|} \operatorname{tr} W^2 \\ &= \| W \| = \| W(C) \| . \end{aligned}$$

Optimal partitions $C_0$ are then the solutions of the problem

$$\| W(C_0) \| = \min_{C \in P^+(n,M)} \| W(C) \| . \tag{5.8}$$

If we apply a class specific norm, for example

$$\| G_j \| = 1 \quad (j=1, \ldots, n), \tag{5.9}$$

and if we optimize (4.4), then we obtain the objective function

$$\min_{C \in P^{++}(n,M)} \sum_{j=1}^{n} \| W_j \| , \tag{5.10}$$

where the $W_j$ denote, as always, the scatter matrices of the $j^{th}$ class.

For neither of the objective functions (5.8) and (5.10) have the properties described for the previous criteria yet seen examined. Whether they have other desirable properties is a question for research.

### 5.2 Other invariant criteria

A criterion $D = D(C)$ is called invariant with respect to certain transformations

$$x_i' = Ax_i + b \quad (i=1, \ldots, m) \tag{5.11}$$

of the given object vectors, provided that the values of the objective functions under both coordinate systems are the same or differ only by a positive constant factor, for every feasible partition.

Since such invariance requirements are often given undue weight in the literature, we wish to discuss a number of criteria which are invariant in the above sense, and to point out their weaknesses. In all cases, we assume the matrix $A \in \mathbb{R}(s,s)$ to be regular.

Since the transformations (5.11) turn all matrices

$$F = W_j, B_j, W, B, T \tag{5.12}$$

into

$$F' = AFA^T \tag{5.13}$$

according to (2.15) to (2.19), all criteria of the form

$$D_F(C) = \sum_{j=1}^{n} \sum_{i \in C_j} (x_i - \bar{x}_j)^T F^{-1} (x_i - \bar{x}_j) \tag{5.14}$$

are invariant with respect to (5.11), provided that F is a linear combination of the matrices (5.12). Since F must be positive definite, only a subset of these linear combinations can be used. Thus, $F = B_j$ is not possible since the rank of $B_j$ equals one, and $B_j$ is thus not invertible for $s > 1$. Since $B = T - W$, we do not look at B either. Because of (4.18), the $W_j$ are similarly not possible. For $F = W$, $D_F(C) = s$ independent of the partition C, as can be shown by calculation (3.20). If we want to avoid arbitrary linear combinations, the only remaining possibility is $F = T$.

Assuming T to be positive definite, there is a Cholesky decomposition

$$T = LL^T \tag{5.15}$$

with a lower triangular matrix L. Using

$$y_i := L^{-1} x_i \quad (i=1, \ldots, m) , \tag{5.16}$$

(5.14) applied to $F = T$ becomes

$$D_T(C) = \sum_{j=1}^{n} \sum_{i \in C_j} \| y_i - \bar{y}_j \|^2 , \tag{5.17}$$

that is, the minimum variance criterion for the object vectors $y_i$ which have been transformed according to (5.16). Once the decomposition (5.15) is achieved and the $y_i$ are calculated from (5.16) using forward substitution, the approximation methods developed for the variance criterion may be applied.

Although this invariant method looks attractively simple, it does have serious drawbacks (see [B18]). The matrix T depends not on the specific clusters sought, but simply on the set of all object vectors and on the overall mean vector. Since, furthermore,

$$\sum_{i=1}^{m} (y_i - \bar{y})(y_i - \bar{y})^T = L^{-1} T L = I \tag{5.18}$$

holds, the object vectors are transformed in such a way that the corresponding total scatter matrix is just the identity matrix. However, such normalization is totally arbitrary.

Other invariant criteria are obtained when we note that W, B, and T transform according to (5.13), and when we remember the following theorem of linear algebra [B2] which relies largely on the simultaneous diagonalization of matrices with certain properties:

**Lemma**: Let $H, G, A \in \mathbb{R}(s,s)$. Let G be positive semi-definite, H positive definite, and A regular. Then the only invariants under the transformations

$$G' = AGA^T, \qquad H' = AHA^T \tag{5.19}$$

are the zeros $\lambda$ of

$$\det(G - \lambda H) = 0 \ , \tag{5.20}$$

that is, the eigenvalues of $H^{-1}G$. Moreover, every $\lambda$ satisfying (5.20) is non-negative.

Of the matrices T, W and B, which are of interest here, T and W are generally positive definite, whereas B is often not. For against the necessary conditions $m > s$ of $m_j > s$ for at least one $j \in \{1, \ldots, n\}$ there is the condition $n > s$. However, B is certainly positive semi-definite. According to the conditions for the above Lemma, the eigenvalues of

$$T^{-1} B$$

$$T^{-1} W = I - T^{-1} B$$

$$W^{-1} B$$

$$W^{-1} T = I + W^{-1} B$$

are suitable candidates for invariants.

If the eigenvalues of $W^{-1}B$ are denoted as $v_k \geqslant 0$, then $1 + v_k$ are the eigenvalues of $W^{-1}T$, $1/(1 + v_k)$ those of $T^{-1}W$, and $v_k/(1 + v_k)$ those of $T^{-1}B$. Hence, we may restrict ourselves to introducing the eigenvalues $v_k$ of $W^{-1}B$.

Any arbitrary, non-constant functions of these eigenvalues [B16], which do depend on the partition C since $W = W(C)$ and $B = B(C)$, are clearly also invariant objective functions [B18]. Let us select from this wide range of possibilities those which are easiest to compute. They are

$$\operatorname{tr}(W^{-1}B) = \sum_{k=1}^{s} v_k \,, \tag{5.21}$$

$$\operatorname{tr}(W^{-1}T) = \sum_{k=1}^{s} (1 + v_k) \,, \tag{5.22}$$

$$\det(W^{-1}B) = \prod_{k=1}^{s} v_k \,, \tag{5.23}$$

$$\det(W^{-1}T) = \prod_{k=1}^{s} (1 + v_k) \,, \tag{5.24}$$

$$\rho(W^{-1}B) = \max_{k=1,\dots,s} v_k \,. \tag{5.25}$$

For the special case of $s = 1$, it has already been shown that minimizing tr W corresponds to maximizing $\operatorname{tr}(W^{-1}B)$. Consequently, the aim must be to maximize the objective functions above over the set of admissible partitions. The latter must always be restricted to $P^{+}(n,M)$ since $W^{-1}$ does not exist otherwise.

Criteria (5.21) and (5.22) are equivalent. The objective function (5.23) cannot generally be used since, at least for $n \leqslant s$, $W^{-1}B$ has at least one eigenvalue equal to zero [B24] for every feasible partition as a result of Theorem 3.1. Because

$$\det(W^{-1}T) = \det / \det W$$

and because det T is constant, maximizing (5.24) is equivalent to minimizing det W [B18]. Finally, objective function (5.25) is hardly feasible in practice because of the amount of computation required, although there is the power method as a simple procedure for finding the largest eigenvalue of $W^{-1}B$. Moreover, properties analogous to those of the minimum variance or determinant criterion are not known or have not been investigated.

For the remaining criteria (5.21) and (5.22) we have

$$\operatorname{tr}(W^{-1}B) = \operatorname{tr}(W^{-1}\sum_{j=1}^{n} m_j (\bar{x}_j - \bar{x})(\bar{x}_j - \bar{x})^T)$$

$$= \sum_{j=1}^{n} m_j \operatorname{tr}(W^{-1}(\bar{x}_j - \bar{x})(\bar{x}_j - \bar{x})^T)$$

$$= \sum_{j=1}^{n} m_j \operatorname{tr}((\bar{x}_j - \bar{x})^T W^{-1} (\bar{x}_j - \bar{x}))$$

$$= \sum_{j=1}^{n} m_j (\bar{x}_j - \bar{x})^T W^{-1} (\bar{x}_j - \bar{x})$$

and similarly

$$\operatorname{tr}(W^{-1}T) = \sum_{i=1}^{m} (x_i - \bar{x})^T W^{-1} (x_i - \bar{x}) .$$

Moreover,

$$\operatorname{tr}(W^{-1}T) = s + \operatorname{tr}(W^{-1}B) .$$

It appears that the maximization of tr $(W^{-1}B)$ corresponds to the maximization of tr B or to the minimization of tr W, using a class-specific transformed Euclidean metric. Since the desirable properties listed at the start of Chapter 3 are largely unknown or unproven for tr $(W^{-1}B)$ and tr $(W^{-1}T)$, it is not advisable to use these criteria merely because of their invariance properties.

**5.3 $L_p$-criteria**

So far, we have only looked at the Euclidean metric and at metrics derived from it through transformations with positive definite matrices G which are dependent on the groups in a partition. Instead, we can use the $L_p$ norm

$$\| x \|_p := \sqrt[p]{\sum_{k=1}^{s} | x_k |^p}, \quad p \geqslant 1, x \in \mathbb{R}^s , \tag{5.26}$$

for measuring distances, and we need not restrict ourselves to using their square $(q = 2)$. In this way we can define the $L_p$ criterion as a generalization of the variance criterion:

$$D_p^{(q)}(C) = \sum_{j=1}^{n} \min_{y_j \in \mathbb{R}^s} \sum_{i \in C_j} \| x_i - y_j \|_p^q , p \geqslant 1, q > 0 . \tag{5.27}$$

Like the variance criterion, $D_p^{(q)}$ is made up from a sum of individual measures of compactness.

The special case with $p = q = 2$ is the variance criterion. This had the pleasant property that the centres $y_j$ arose simply as the class mean vectors.

Another simple case is $p = q = 1$. We refer to

$$D_1^{(1)}(C) = \sum_{j=1}^{n} \min_{y_j \in \mathbb{R}^s} \sum_{i \in C_j} \| x_i - y_j \|_1 \tag{5.28}$$

as the $L_1$ criterion [B39]. Here, as for $D_\infty^{(1)}$, the determination of the centres $y_j$ is also very simple. We shall return to this in Chapter 6.

For other values of p and q, the determination of an optimal partition $C_0$ according to

$$D_p^{(q)}(C_0) = \min_{C \in P(n,M)} D_p^{(q)}(C) \tag{5.29}$$

is a mixed problem of continuous and combinatorial optimization which involves a highly non-linear determination of centres. The latter has to be done iteratively and thus prevents the use of explicit up- and downdating formulae. For $p = 2$ and $q = 1$, a program for the corresponding location–allocation problem can be found in [A20].

This statement also applies when (5.27) is generalized in analogy with the criterion of adaptive distances to

$$D_{p_1,\ldots,p_n}^{(q_1,\ldots,q_n)} = \sum_{j=1}^{n} \min_{y_j \in \mathbb{R}^s} \sum_{i \in C_j} \| x_i - y_j \|_{p_j}^{q_j}, \tag{5.30}$$

$$p_j \geqslant 1, q_j > 0 \quad (j=1, \ldots, n).$$

The problem of being unable to calculate centres explicitly (unless $p_j = q_j = 2$ or $p_j = q_j = 1$ or $p_j = \infty$ and $q_j = 1$) is the reason why criteria (5.27) and (5.30) have not been considered so far. There is practically no chance of being able to find properties corresponding to those of the criteria discussed in Chapters 2 to 4.

## 5.4 The use of generalized means

In all the objective functions considered, with the exception of (4.15), the task was one of minimizing a sum of cluster specific measures of compactness. For the variance criterion these were

$$e(C_j) = \operatorname{tr} W_j = \sum_{i \in C_j} \| x_i - \bar{x}_j \|^2,$$

for the criterion of adaptive distances they were

$$e(C_j) = (\det W_j)^{1/s},$$

and for the $L_p$ criterion

$$e(C_j) = \min_{y_j \in \mathbb{R}^s} \sum_{i \in C_j} \| x_i - y_j \|_p^q.$$

In the case of the determinant criterion,

$$e(C_j) = \sum_{i \in C_j} (x_i - \bar{x}_j)^T (\det W)^{1/s} W^{-1} (x_i - \bar{x}_j)$$

was also dependent on the partition as a whole, since $W = W(C)$.

In all cases, we have minimized the sum of relevant measures of compactness, or, equivalently, their mean

$$\frac{1}{n} \sum_{j=1}^{n} e(C_j) .$$

It is apparent that more generalized ways of calculating means could be used, as we have already done in (4.15).

Let

$$z_j \geqslant 0, \; h_j > 0 \; (j=1, \ldots, n), \; h := \sum_{j=1}^{n} h_j, \; \beta \in \mathbb{R}. \tag{5.31}$$

Then

$$M_\beta(h,z) = \sqrt[\beta]{\frac{1}{h} \sum_{j=1}^{n} h_j z_j^\beta} \tag{5.32}$$

is known as the generalized mean of order $\beta$ [B34]. For $z_j \geqslant 0$ it is defined for $\beta \in (0,\infty]$, and for $z_j > 0$ it is defined for $\beta \in [-\infty,\infty]$.

When $z_j > 0$ ($j=1, \ldots, n$) and $\beta = -1$, we have the weighted harmonic mean

$$M_{-1}(h,z) = h \Big/ \sum_{j=1}^{n} (h_j / z_j) , \tag{5.33}$$

when $\beta = 0$ the weighted geometric mean

$$M_0(h,z) = \Big( \prod_{j=1}^{n} z_j^{h_j} \Big)^{1/h} , \tag{5.34}$$

and when $\beta = -\infty$

$$M_{-\infty}(h,z) = \min_{j=1,\ldots,n} z_j . \tag{5.35}$$

For $z_j \geqslant 0$ and $\beta = \infty$,

$$M_\infty(h,z) = \max_{j=1,\ldots,n} z_j , \tag{5.36}$$

and when $\beta = 1$, we have the weighted arithmetic mean

$$M_1(h,z) = \frac{1}{h} \sum_{j=1}^{n} h_j z_j , \tag{5.37}$$

which reduces to the arithmetic mean as we have used it so for when $h_j = 1$ ($j=1, \ldots, n$).

More generally than before, we can consider objective functions of the form

$$\sqrt[\beta]{\frac{1}{h} \sum_{j=1}^{n} h_j (e(C_j))^\beta} \tag{5.38}$$

as criteria for cluster dissection. For the minimum variance, determinant, and $L_p$ criteria $e(C_j) \geqslant 0$. Consequently, we need to restrict $\beta$ to $(0, \infty]$. With the criterion of adaptive distances, $e(C_j) > 0$ since $C \in P^{++}(n,M)$, and we can consequently allow $\beta \in [-\infty, \infty]$. For example, with $h_j = 1$ $(j=1, \ldots, n)$ we obtain the objective functions

$$\min_{C \in P(n,M)} \sqrt[\beta]{\frac{1}{n} \sum_{j=1}^{n} (\text{tr } W_j)^\beta}, \quad \beta > 0 \tag{5.39}$$

and

$$\min_{C \in P^{++}(n,M)} \sqrt[\beta]{\frac{1}{n} \sum_{j=1}^{n} (\det W_j)^\beta}, \quad \beta \in [-\infty, \infty]. \tag{5.40}$$

For $\beta > 0$, the objective functions

$$\min_{C \in P(n,M)} \sum_{j=1}^{n} (\text{tr } W_j)^\beta, \quad \beta > 0, \tag{5.41}$$

and

$$\min_{C \subset P^{++}(n,M)} \sum_{j=1}^{n} (\det W_j)^\beta, \quad \beta > 0, \tag{5.42}$$

are equivalent to (5.39) and (5.40). The special case of (5.41) with $\beta = 1$ is the variance criterion. With $\beta = 1$, (5.42) becomes the objective function considered in (4.15), and with $\beta = 1/s$ it becomes the criterion of adaptive distances. It follows from (5.34) that, for $\beta = 0$, the objective function is equivalent to

$$\min_{C \in P^{++}(n,M)} \prod_{j=1}^{n} (\det W_j). \tag{5.43}$$

For $\beta < 0$,

$$\max_{C \in P^{++}(n,M)} \sum_{j=1}^{n} (\det W_j)^\beta, \quad \beta < 0 \tag{5.44}$$

is equivalent to (5.40). The restriction to positive values of $\beta$ which was required in (4.15) makes it clear that maximization is required here.

If, on the other hand, we set $h_j = m_j$ and $z_j = \det W_j/m_j$ $(j=1, \ldots, n)$ then we obtain for $\beta = 0$

$$\min_{C \in P^{++}(n,M)} \prod_{j=1}^{n} \left(\frac{1}{m_j} \det W_j\right)^{m_j}. \tag{5.45}$$

A corresponding objective function for $z_j = \operatorname{tr} W_j/m_j$ is not sensible since $\operatorname{tr} W_j \geqslant 0$. Other values of $h_j$ may also be considered.

Finally, $\beta = \infty$ yields the objective function

$$\min_{C \in P(n,M)} \max_{j=1,\ldots,n} \operatorname{tr} W_j \tag{5.46}$$

or

$$\min_{C \in P^{++}(n,M)} \max_{j=1,\ldots,n} \det W_j. \tag{5.47}$$

Although all these objective functions could be used in principle, it is necessary to clarify first whether certain properties, either known from the variance criterion or new ones, are present. For example, we have seen in (5.43) that $\beta$ needs to be restricted to $0 < \beta \leqslant 1$, otherwise the minimal distance property cannot be proved to be a necessary condition for optimal partitions.

## 5.5 Side conditions

Independent of the objective function chosen, certain ideas about the composition of clusters may exist; less where a structure is being analysed, more where a suitable division into classes is sought. Consequently, explicit constraints may be imposed on the clusters produced.

For example, it may not be desirable to admit all partitions $C \in P(n,M)$ with $m_j > 0$. Instead, the number of elements in a cluster may be limited to

$$0 < m_0 \leqslant m_j \leqslant m_u \quad (j=1, \ldots, n) \tag{5.48}$$

Here, sensible values for the integers $m_0$ and $m_u$ must be chosen, that is,

$$m_0 n \leqslant m \leqslant m_u n \tag{5.49}$$

must hold. It is also conceivable that (5.48) be demanded for some classes only, or that class-specific values are specified in advance, and (5.49) altered accordingly. In all cases, the exchange method can easily be modified to ensure that, starting from a partition satisfying the above constraints, only those exchange steps are considered which do not change the feasibility of the partition. In the implementation of the exchange method for the variance criterion we have allowed for

$$m_j \geqslant m_0 > 0. \tag{5.50}$$

The value $m_0 = 1$ corresponds to the normal case.

Side conditions may also take an entirely different form. For $e(C_j) = \text{tr } W_j$ as a measure of compactness, there may exist other conditions in addition to (5.48) or instead of it, of the form

$$0 \leqslant d_0 \leqslant \| x_i - \bar{x}_j \| \leqslant d_u, \quad i \in C_j, \tag{5.51}$$

$$0 \leqslant e_0 \leqslant e(C_j) \leqslant e_u, \quad j=1, \ldots, n, \tag{5.52}$$

$$0 \leqslant f_0 \leqslant \| \bar{x}_j - \bar{x}_k \| \leqslant f_u, \quad j \neq k. \tag{5.53}$$

These may apply in a class-specific way or relative to other measures of homogeneity and to the corresponding class centres $y_j$. Of course, the restrictions would have to be such that feasible partitions with the right properties do exist. The more restrictive the side conditions, the smaller is the number of feasible partitions. This does not, however, make the task of finding an optimal partition among the feasible ones any easier. The opposite is the case, since desirable properties such as the minimal distance property as a necessary condition for optimality or the separation property are, in general, destroyed.

For the greater variety of conceivable approximation methods, the main difficulty lies in finding feasible starting partitions, or in proving their existence. After that, one can apply the minimal distance or exchange method while maintaining feasibility, and try to reduce the objective function. There also exist a number of not very promising attempts (see [B55]) at introducing penalty functions as they are used with continuous optimization. These allow starting partitions which are not themselves feasible. A very simple, but usually not very successful, method consists of taking end partitions from the exchange method, say, and of changing these end partitions until they conform with the conditions imposed, while increasing the value of the objective function as little as possible.

The most exotic side conditions which the author ever encountered in a practical problem were as follows. With each object vector $x_i \in \mathbb{R}^s$ there was associated a vector $z_j \in \mathbb{R}^t$ of secondary variables which were not, however, to be used for the classification. Instead, the variance criterion was to be minimized with respect to those partitions for which the mean vectors $\bar{z}_j$ of the secondary variables conformed to the restriction

$$a_j \leqslant \bar{z}_j \leqslant b_j, \quad a_j, b_j \in \mathbb{R}^t \quad (j=1, \ldots, n) \tag{5.54}$$

The significance of the problems outlined here, especially for the purposes of cluster dissection, should not be underestimated. They are still largely unresearched.

**Problem 5.1** (continuation of problem 3.2): Show that, for $n = 2$, minimizing $\det W$ is equivalent to maximizing $\text{tr}\,(W^{-1}B)$.

**Problem 5.2**: Develop exchange formulas for the following objective functions:

(a) $\max_{C \in P(n,M)} \operatorname{tr}(W^{-1}B)$ or $\max_{C \in P(n,M)} \operatorname{tr}(W^{-1}T)$,

(b) $\min_{C \in P(n,M)} \sum_{j=1}^{n} (\operatorname{tr} W_j)^{\beta}$, $\beta > 0$,

(c) $\min_{C \in P^{+}(n,M)} \| W \|$

(d) $\min_{C \in P^{++}(n,M)} \sum_{j=1}^{n} \| W_j \|$.

In which cases can properties like those possessed by the minimum variance criterion be proven? What invariance properties hold?

**Problem 5.3**: Formulate a penalty function for the variance criterion and side conditions of the form (5.51), (5.52) or (5.53). Can the exchange method be modified to take account of this?

# 6 The $L_1$ criteria for quantitative, binary, and ordinal data

## 6.1 The $L_1$-Criterion and the use of median vectors as centres

With the variance criterion, the quantity

$$e(C_j) = \min_{y \in \mathbb{R}^s} \sum_{i \in C_j} \| x_i - y \|^2 = \sum_{i \in C_j} \| x_i - \bar{x}_j \|^2 = S(\bar{x}_j)$$

was used as the measure of compactness for a cluster.

Now let us replace the sum of squared deviations from the corresponding centre $\bar{x}_j$ by the sum of absolute values of deviations from a new centre y to be defined later (see [B39]). Using the notation

$$\| x \|_1 := \sum_{k=1}^{s} | x_k |, \quad x \in \mathbb{R}^s$$

we need to minimize the function

$$T(y) = \sum_{i \in C_j} \| x_i - y \|_1 = \sum_{i \in C_j} \left( \sum_{k=1}^{s} | x_{ik} - y_k | \right)$$

$$= \sum_{k=1}^{s} \left( \sum_{i \in C_j} | x_{ik} - y_k | \right). \tag{6.1}$$

In analogy to the variance criterion, we can define and examine the $L_1$-criterion

$$D_1(C) = \sum_{j=1}^{n} \min_{y_j \in \mathbb{R}^s} \sum_{i \in C_j} \| x_i - y_j \|_1 . \tag{6.2}$$

Because of (6.1), T(y) may be written in the form

$$T(y) = \sum_{k=1}^{s} t(y_k) \tag{6.3}$$

where

$$t(y_k) = \sum_{i \in C_j} | x_{ik} - y_k | \quad (k=1, \ldots, s) \; . \tag{6.4}$$

Consequently, $T = T(y)$ may be minimized by minimizing the functions $t = t(y_k)$ separately. If we omit the subscript k for the moment, our task is one of the minimizing s function of the form

$$t(y) = \sum_{i \in C_j} | x_i - y | , \quad x_i \in \mathbb{R}, y \in \mathbb{R} , \tag{6.5}$$

independently by each other. As a sum of continuous functions of y, $t = t(y)$ is continuous. We shall show that it forms a convex polygon. For this, we sort the $x_i$ according to size. For the sorted numbers $\hat{x}_i$ we have

$$\hat{x}_1 \leqslant \hat{x}_2 \leqslant \ldots \leqslant \hat{x}_{m_j} \tag{6.6}$$

and also

$$t(y) = \sum_{i \in C_j} | \hat{x}_i - y | . \tag{6.7}$$

In each of the different intervals given by (6.6), (6.7) may be rewritten in the form

$$t(y) = \sum_{i=1}^{k} (y - \hat{x}_i) + \sum_{i=k+1}^{m_j} (\hat{x}_i - y), \quad y \in [\hat{x}_k, \hat{x}_{k+1}] , \tag{6.8}$$

$$k = 1, \ldots, m_j - 1.$$

Thus in the interval $[\hat{x}_1, \hat{x}_{mj}]$, the function $t = t(y)$ consists of straight lines, continuously joined at $x_k (k=2, \ldots, m_j - 1)$. As may be seen from (6.8), the lines have the slopes

$$k - (m_j - k), \quad (k=1, \ldots, m_j - 1), \tag{6.9}$$

more specifically, the slopes are

$$- m_j + 2, - m_j + 4, \ldots, - 1, 1, 3, \ldots, m_j - 2, \text{ when } m_j \text{ is odd}$$

and

$$- m_j + 2, - m_j + 4, \ldots, - 2, 0, 2, \ldots, m_j - 2, \text{ when } m_j \text{ is even} \tag{6.10}$$

Because of its continuity, $t = t(y)$ is thus proved to be a convex polygon. It follows from the values of the slopes in (6.10) that the desired value of y which minmizes $t = t(y)$ is

$$y = \hat{x}_g \text{ when } m_j \text{ is odd and } g := (m_j + 1)/2,$$
$$y = [\hat{x}_g, \hat{x}_{g+1}]\text{, when } m_j \text{ is even and } g := m_j/2 \ . \tag{6.11}$$

These values are known as the median of the numbers $x_i \in \mathbb{R}$, $i \in C_j$. For odd $m_j$, the median is uniquely defined; for even $m_j$ it consists of a whole interval which only shrinks to a single point when $\hat{x}_g = \hat{x}_{g+1}$. The calculation of the median is simple. The numbers $x_i$ are sorted by size, and then the values corresponding to (6.11) are extracted from (6.6) via the index g.

If (6.4) is solved for each component a vector $y = (y_1 \ldots, y_s)^T$ is obtained which minimizes (6.1). We call it the median vector. When $m_j$ is even, its components may consist of intervals.

Clearly, the value or range of the median depends only on the ranking of the $x_i$. In contrast to the mean, its value is therefore not affected by outliers. Where an ordinal variable is represented by $1, \ldots, t$, the differences between successive categories may well not have an equal size or significance. In this situation, the median is unaffected, whereas the mean is. Furthermore, in the measure of compactness

$$e(C_j) = \min_{y_j \in \mathbb{R}^s} \sum_{i \in C_j} \| x_i - y_j \|_1 \tag{6.12}$$

which is a sum of absolute deviations, outliers in the data are given less prominence than in the sum of squared deviations. For that reason in particular, the objective function (6.2) is suitable for use with an ordinal data matrix. It is also used where the formation of absolute differences seems preferable to that of squared differences, as in the dissection of time series (see [B44]).

The advantages of the $L_1$-norm mentioned above make it preferable to the squared $L_2$-norm of the difference between vectors in certain situations. By comparison, the $L_\infty$-norm

$$\| x \|_\infty := \max_{k=1,\ldots,s} | x_k |, \quad x \in \mathbb{R}^s,$$

is even less suitable. The reason is that the function

$$U(y) = \sum_{i \in C_j} \| x_i - y \|_\infty = \sum_{i \in C_j} \max_{k=1,\ldots,s} | x_{ik} - y_k | \tag{6.13}$$

is minimized by the vector y with components

$$y_k = \frac{1}{2} (\max_{i \in C_j} x_{ik} - \min_{i \in C_j} x_{ik}), \tag{6.14}$$

These values are determined solely by $\hat{x}_{1,k}$ and $\hat{x}_{m_j,k}$ and are therefore extremely sensitive to outliers in the data. For this reason, the objective function

$$D_\infty(C) = \sum_{j=1}^{n} \min_{y_j \in \mathbb{R}^s} \sum_{i \in C_j} \| x_i - y_j \|_\infty \tag{6.15}$$

is unsuitable for practical applications. Only the properties of $D = D_1(C)$ defined in (6.2) will therefore be examined below.

## 6.2 Invariance properties

If we add a constant to the numbers $\hat{x}_i$ in (6.6), the chain of inequalities is retained. The value of the index g is unaffected. Thus, our objective function (6.2) is invariant with respect to translations $x_i' = x_i + b$ $(i=1, \ldots, m)$ of the object vectors.

The following theorem states which matrices $A \in \mathbb{R}(s,s)$ and which transformations $x_i' = Ax_i$ leave the $L_1$- and $L_\infty$-norms, and hence our objective function (6.2), invariant.

**Theorem 6.1**: Precisely those non-singular matrices $A \in \mathbb{R}(s,s)$ leave the $L_1$- and the $L_\infty$-norms invariant, that interchange components of $x \in \mathbb{R}^s$ and/or change their sign.

*Proof:* Because of

$$\| x \|_1 = \sum_{k=1}^{s} | x_k |, \quad \| x \|_\infty = \max_{k=1,\ldots,s} | x_k |$$

it is clear that matrices with the stated properties do not change the norms. Conversely, let

$$\| Ax \|_\infty = \| x \|_\infty \text{ for all } x \in \mathbb{R}^s. \tag{6.16}$$

The matrix norm induced by the vector norm $\| \,.\, \|_\infty$ is (see [B54])

$$\| A \|_\infty := \max_{x \neq 0} \frac{\| Ax \|_\infty}{\| x \|_\infty} = \max_{i=1,\ldots,s} \sum_{k=1}^{s} | a_{ik} | . \tag{6.17}$$

It follows from assumption (6.16) that

$$\max_{i=1,\ldots,s} \sum_{k=1}^{s} | a_{ik} | = 1 , \tag{6.18}$$

and hence that

$$\sum_{k=1}^{s} | a_{ik} | \leqslant 1 \quad (i=1, \ldots, s). \tag{6.19}$$

Now substitute for x in (6.16) the $k^{th}$ unit vector $e_k$ in $\mathbb{R}^s$. We get

$$1 = \| e_k \|_\infty = \| Ae_k \|_\infty = \max_{i=1,\ldots,s} | a_{ik} | . \tag{6.20}$$

Thus, for each k = 1, . . ., s there exists a p with

$$| a_{pk} | = 1. \tag{6.21}$$

Hence,

$$a_{pk} = \pm 1, \tag{6.22}$$

and it follows from (6.19) that

$$a_{pj} = 0 \text{ for } j \neq k. \tag{6.23}$$

On the other hand, for each k there is at least one p for which (6.21) holds. For if there existed a $q \neq p$ with $| a_{qk} | = 1$, then, by the same reasoning, $a_{qk} = \pm 1$ and $a_{qj} = 0$ for $j \neq k$. Together, we would then have

$$a_{pk} = \pm a_{qk},\ a_{pj} = a_{qj} = 0 \text{ für } j \neq k. \tag{6.24}$$

This would contradict the non-singularity of A, since two rows would be linearly dependent. Thus, there exists for each k only one p with (6.22) and (6.23). Moreover, since A is regular, the values of p corresponding to different values of k must be different. This proves that, for the $L_\infty$-norm, A is of the form specified in the theorem. But it is well known (see [B54]) that

$$\| A \|_1 = \| A^T \|_\infty = \max_{k=1,\ldots,s} \sum_{i=1}^{s} | a_{ik} | . \tag{6.25}$$

Since $A^T$ has the same form as A, the assertion of the theorem also holds for the $L_1$-norm.

Thus, the value of our objective function (6.2) is unchanged precisely in those instances when components of the given vectors $x_i$ are interchanged, or when their sign is changed, or when a constant vector b is added. Applying Theorem 6.1 to the special case of scale transformations (2.29), we obtain the same result as for the variance criterion.

**Theorem 6.2**: The $L_1$-criterion is invariant with respect to transformations of scale (2.29) in precisely those cases when $h_k = \pm 1$ (k = 1, . . ., s), that is, it is only invariant with respect to changes in sign.

### 6.3 Minimal distance method

As before, we define optimal partitions $C_0$ for the $L_1$-criterion by

$$D_1(C_0) = \min_{C \in P(n,M)} D_1(C) \quad . \tag{6.26}$$

If we wish to define a minimal distance partition as in (2.45), we must overcome the difficulty that the median vector $y_j$ may not be uniquely defined when the number $m_j$ is even. In this context, we therefore define a fixed vector $\tilde{y}_j$ whose components are

$$\begin{aligned} \tilde{y}_{jk} &= y_{jk} = \hat{x}_{gk} && (m_j \text{ odd}) \\ \tilde{y}_{jk} &= \frac{1}{2} (\hat{x}_{gk} + \hat{x}_{g+1,k}) && (m_j \text{ even}) \; . \end{aligned} \tag{6.27}$$

Then we define a minimal distance partition as usual, by

$$i \in C_j \Longleftrightarrow \| x_i - \tilde{y}_j \|_1 = \min_{k=1,\ldots,n} \| x_i - \tilde{y}_k \|_1 \; . \tag{6.28}$$

To modify the minimal distance method in section 2.7 for the $L_1$-criterion, we need to replace the computation of the mean vector (2.46) by that of the median vector in its modified form (6.27), and in (2.47) we need to apply definition (6.28). Then we have the following theorem.

**Theorem 6.3**: In applying the minimal distance method with the modifications just described, the $L_1$-criterion does not increase.

*Proof:*

$$\begin{aligned} D_1(C^{(t)}) &= \sum_{j=1}^{n} \sum_{i \in C_j^{(t)}} \| x_i - \tilde{y}_j^{(t)} \|_1 \\ &\geqslant \sum_{j=1}^{n} \sum_{i \in C_j^{(t)}} \min_{k=1,\ldots,s} \| x_i - \tilde{y}_j^{(t)} \|_1 \\ &= \sum_{j=1}^{n} \sum_{i \in C_j^{(t+1)}} \| x_i - \tilde{y}_j^{(t)} \|_1 \\ &\geqslant \sum_{j=1}^{n} \sum_{i \in C_j^{(t+1)}} \| x_i - \tilde{y}_j^{(t+1)} \|_1 \\ &= D_1(C^{(t+1)}). \end{aligned}$$

As always, we used the minimal distance property and the property that the class centre minimizes $T(y)$ according to (6.1), in applying estimates.

## 6.4 Up- and downdating formulae for median vectors, exchange method, and a necessary condition for the optimality of partitions

Because of (6.1), we can rewrite the $L_1$-criterion (6.2) in the form

$$D_1(C) = \sum_{j=1}^{n} \left( \sum_{k=1}^{s} t_k(C_j) \right) \tag{6.29}$$

with

$$t_k(C_j) = \min_{y_{jk} \in \mathbb{R}} \sum_{i \in C_j} | x_{ik} - y_{jk} | . \tag{6.30}$$

For the exchange method we need to derive formulas showing how the $k^{th}$ component ($k=1, \ldots, s$) of the $j^{th}$ median vector ($j=1, \ldots, n$) changes, and how the component-specific measures of compactness $t_k(C_j)$ change when an object $i \in C_p$ with $m_p > 1$ is taken out of its cluster and moved into a cluster $C_j$ with $j \neq p$.

Although the formulae will turn out to be straightforward, many different cases will have to be distinguished. To keep things simple, we shall, for the time being, suppress as many subscripts as possible and consider the following situation.

Given $x_1, \ldots, x_m \in \mathbb{R}$, let us denote by $\hat{x}_1, \ldots, \hat{x}_m$ the same numbers, but ordered by size. For even m, let $g = m/2$, and for odd m let $g = (m+1)/2$. Then $y = \hat{x}_g$ or $y = [\hat{x}_g, \hat{x}_{g+1}]$ is the median of $x_1, \ldots, x_m$. Use $y'$ to denote the median of the numbers $x_1, \ldots, x_m$ $(m>1)$ with $x_i$ omitted, and $y''$ for the median of $x_1, \ldots, x_{m+1}$ where, without loss of generality, we have set $i = m+1$. Finally, let the corresponding sums of absolute deviations from the median be denoted by

$$t(y) = \sum_{h=1}^{m} | x_h - y | , \tag{6.31}$$

$$t(y') = \sum_{\substack{h=1 \\ h \neq i}}^{m} | x_h - y' | , \tag{6.32}$$

$$t(y'') = \sum_{h=1}^{m+1} | x_h - y'' | . \tag{6.33}$$

Our task is to express $y'$ and $t(y')$ as well as $y''$ and $t(y'')$ as functions of y and $t(y)$ in the simplest possible way.

The following up- and downdating formulae may be formally derived from notations such as (6.8), but are more easily visualized using the line of real numbers.

For removing $x_i$ and odd values of $m \geqslant 3$ we have

$$\begin{aligned} &x_i < y \Rightarrow y' = [\hat{x}_g, \hat{x}_{g+1}] \,, \\ &x_i = y \Rightarrow y' = [\hat{x}_{g-1}, \hat{x}_{g+1}] \,, \\ &x_i > y \Rightarrow y' = [\hat{x}_{g-1}, \hat{x}_g] \,, \end{aligned} \tag{6.34}$$

and in all cases

$$t(y') = t(y) - | x_i - y | \,. \tag{6.35}$$

For even values of $m \geqslant 2$ ($y = [\hat{x}_g, \hat{x}_{g+1}]$) we have

$$\begin{aligned} &x_i \leqslant \hat{x}_g \Rightarrow y' = \hat{x}_{g+1} \,, \\ &x_i > \hat{x}_g \Rightarrow y' = \hat{x}_g \,, \end{aligned} \tag{6.36}$$

and in any case,

$$t(y') = t(y) - | x_i - y' | \,. \tag{6.37}$$

When adding $x_i$ when $m = 1$ we have

$$\begin{aligned} &x_i \leqslant y \Rightarrow y'' = [x_i, y] \,, \\ &x_i > y \Rightarrow y'' = [y, x_i] \,, \end{aligned} \tag{6.38}$$

and for odd values of $m \geqslant 3$,

$$\begin{aligned} &x_i \leqslant y \Rightarrow y'' = [\max(\hat{x}_{g-1}, x_i), \hat{x}_g] \,, \\ &x_i > y \Rightarrow y'' = [\hat{x}_g, \min(\hat{x}_{g+1}, x_i)] \,, \end{aligned} \tag{6.39}$$

and in all cases

$$t(y'') = t(y) + | x_i - y | \,. \tag{6.40}$$

For even values of $m \geqslant 2$ we have

$$\begin{aligned} &x_i \leqslant \hat{x}_g &&\Rightarrow y'' = \hat{x}_g \,, \\ &\hat{x}_g < x_i < \hat{x}_{g+1} &&\Rightarrow y'' = x_i \,, \\ &x_i \geqslant \hat{x}_{g+1} &&\Rightarrow y'' = \hat{x}_{g+1} \,, \end{aligned} \tag{6.41}$$

and again in all cases

$$t(y'') = t(y) + | x_i - y'' | \,. \tag{6.42}$$

For the attempted exchange involving $C_p' := C_p \backslash \{i\}$ and $C_j' := C_j \cup \{i\}$ for $i \in C_p$, with $m_p \geqslant 1$ and $j \neq p$, we need to calculate the medians and values $t_k(C_p)$ and

$t_k(C_j)$ component-wise. For each component, the different cases described above need to be distinguished. Because of the additive structure of the $L_1$-norm, and because (6.35), (6.37), (6.40), and (6.42) only involve componentwise subtraction or addition, the changes in the measure of compactness can be summarized as

$$e(C'_p) = e(C_p) - \| x_i - y_p \|_1, \quad m_p \text{ odd}, \tag{6.43}$$

$$e(C'_p) = e(C_p) - \| x_i - y'_p \|_1, \quad m_p \text{ even}, \tag{6.44}$$

or

$$e(C'_j) = e(C_j) + \| x_i - y_j \|_1, \quad m_j \text{ odd}, \tag{6.45}$$

$$e(C'_j) = e(C_j) + \| x_i - y''_j \|_1, \quad m_j \text{ even} . \tag{6.46}$$

Here $y'_p$ denotes the median vector of $C'_p$, and $y''_j$ the median vector of $C'_j$.

For the moment, let $\tilde{y}_p$, $\tilde{y}_j$, $\tilde{y}'_p$, and $\tilde{y}''_j$ denote the centres of the median intervals of $C_p$, $C_j$, $C'_p$ and $C'_j$ respectively when the number of elements is even, and let them equal $y_p$, $y_j$, $y'_p$, and $y''_j$ resp. when that number is odd. For each component, (6.34) and (6.36) apply, and summing over all components gives

$$\| x_i - \tilde{y}'_p \|_1 \geqslant \| x_i - \tilde{y}_p \|_1 . \tag{6.47}$$

Similarly, (6.38), (6.39) and (6.41) imply

$$\| x_i - \tilde{y}_j \|_1 \geqslant \| x_i - \tilde{y}''_j \|_1 . \tag{6.48}$$

Now we can demonstrate the following theorem.

**Theorem 6.4**: Using the $L_1$-criterion, every optimal partition is a minimal distance partition in the sense of (6.28).

*Proof:* If the minimal distance property did not hold, there would exist an $i \in C_p$ with $m_p > 1$ and a $j \neq p$ with

$$\| x_i - \tilde{y}_p \|_1 > \| x_i - \tilde{y}_j \|_1 . \tag{6.49}$$

If $m_p$ and $m_j$ are both odd, (6.43) and (6.45) can be applied to obtain

$$e(C_p) - e(C'_p) > e(C'_j) - e(C_j) . \tag{6.50}$$

If $m_p$ or $m_j$ is even, the left hand side of (6.49) may be increased using (6.47), and the right hand side may be decreased using (6.48). Applying (6.44) or (6.46) similarly leads to (6.50).

If (6.50) is rewritten as

$$e(C'_p) + e(C'_j) < e(C_p) + e(C_j) ,$$

it becomes clear that a successful exchange, reducing the value of the objective function, would be possible. This, however, contradicts the optimality of the partition.

## 6.5 A counter example to separating hyperplanes further properties

The $L_1$-criterion has much in common with the variance criterion: simply calculated centres; simple if confusing up- and downdating formulae for median vectors and measures of compactness (due to the need to distinguish special cases); Theorems 6.2, 6.3 and 6.4. We shall now point out some properties distinguishing the two criteria.

First, there is no decomposition analogous to $\text{tr}\, T = \text{tr}\, W + \text{tr}\, B$, and hence no theorem analogous to 2.2. The reason is, of course, that the $L_1$-norm is not induced by a scalar product. Neither is it possible to eliminate the median vectors from the objective function in the way that the mean vectors were eliminated from tr W. Simple counter examples can be used to show that no theorem like Theorem 2.3 holds for the $L_1$ criterion.

Neither is there a separation theorem like Theorem 2.9. The reason is that $\{x : \|x\|_1 \leqslant 1\}$ is not a sphere like $\{x : \|x\| \leqslant 1\}$, but a complicated, not strictly convex, shape which is a square in $\mathbb{R}^2$ and an octahedron in $\mathbb{R}^3$. The set of all points $x \in \mathbb{R}^s$ which have the same $L_1$-distance from two points, for example from the cluster medians $y_p$ and $y_q$; that is,

$$\{x : \|x - y_p\|_1 = \|x - y_q\|_1\},$$

is usually not a hyperplane. For consider the points $x_i$ ($i=1, \ldots, 10$) in $\mathbb{R}^2$, whose Cartesian coordinates are given by

$$\begin{pmatrix}0\\-2\end{pmatrix}, \begin{pmatrix}-2\\0\end{pmatrix}, \begin{pmatrix}0\\0\end{pmatrix}, \begin{pmatrix}0\\10\end{pmatrix}, \begin{pmatrix}5\\0\end{pmatrix}, \begin{pmatrix}7\\4\end{pmatrix}, \begin{pmatrix}8\\5\end{pmatrix}, \begin{pmatrix}8\\4\end{pmatrix}, \begin{pmatrix}8\\3\end{pmatrix}, \begin{pmatrix}9\\4\end{pmatrix}$$

Clearly, $C_1 = \{1, \ldots, 5\}$ and $C_2 = \{6, \ldots, 10\}$ is an optimal partition of length 2. The median vectors, which are unique in this case, are $y_1 = x_3$ and $y_2 = x_8$, as may be seen in Fig. 6.1. The solid line in the figure marks the points which are equidistant from the two medians. Point $x_{11} \coloneqq (3, 4)^T$ lies on the dotted line connecting $x_4$ and $x_5$, both of which belong to $C_1$. But $x_{11}$ has a smaller $L_1$-distance from $y_2$ than from $y_1$.

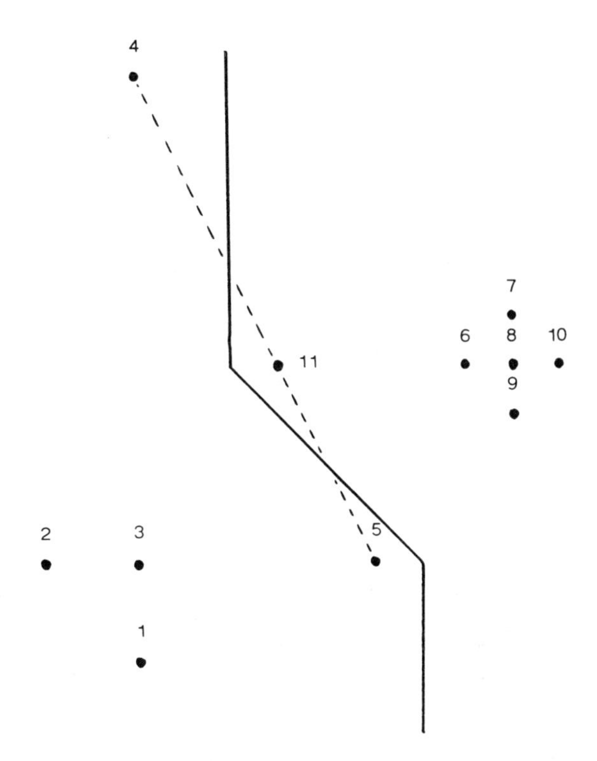


Fig. 6.1

## 6.6 The exchange method for a binary data matrix

In this section we examine the $L_1$-criterion specifically for binary vectors†

$$x_i \in \mathbb{B}^s, \quad \mathbb{B} = \{0, 1\}, \quad (i=1, \ldots, m). \tag{6.51}$$

Its use with a binary data matrix is appropriate whenever zero and one have the same significance for the aim of the cluster dissection or analysis. The criterion and the exchange method can then be greatly simplified. Trivially, any variable with more than $l > 2$ unordered values (that is, a nominal variable) may be transformed into $l$ binary variables.

For the moment, let us omit the subscript k and sometimes j as well. Then (6.30) becomes

$$t(C_j) = \min_{y \in \mathbb{R}} \sum_{i \in C_j} |x_i - y|. \tag{6.52}$$

Since the $x_i$ and the median y can only take on the values zero and one, we have $y \in \mathbb{B}$, and hence

$$t(C_j) = \min \Big( \sum_{i \in C_j} |x_i|, \sum_{i \in C_j} |x_i - 1| \Big)$$

† The set of binary 'vectors' does not form a vector space.

$$= \min \left( \sum_{i \in C_j} x_i, m_j - \sum_{i \in C_j} x_i \right)$$

$$= \min (a_j, m_j - a_j) , \tag{6.53}$$

where $a_j$ denotes the number of ones, and $m_j - a_j$ the number of zeros among the $x_i$ $(i=1, \ldots, m)$.

For the median y which minimizes (6.52) we can apply the same notation:

$$m_j \text{ odd} \Rightarrow y = \begin{cases} 0 \text{ for } a_j < m_j - a_j \\ 1 \text{ for } a_j > m_j - a_j , \end{cases}$$
$$m_j \text{ even} \Rightarrow y = \begin{cases} 0 \text{ for } & a_j < m_j - a_j \\ [0,1] \text{ for } & a_j = m_j - a_j \\ 1 \text{ for } & a_j > m_j - a_j . \end{cases} \tag{6.54}$$

Since the median y does not appear in (6.53), the objective function (6.29) can be rewritten in the much simpler form

$$D_1(C) = \sum_{j=1}^{n} \sum_{k=1}^{s} \min (a_{jk}, m_j - a_{jk}) \tag{6.55}$$

where $a_{jk}$ denotes the number of ones among the $x_{ik}$ $(i \in C_j, k=1, \ldots, s)$. Once an optimal partition for (6.55) or an approximate solution has been found, the components of the medians can, if necessary, be determined using (6.54).

According to (6.55), the $L_1$-criterion for binary vectors means that clusters are simultaneously sought such that for each component of the vectors in a cluster as many zeros or ones occur in such a way that the total value over all classes is minimized. The invariance property of section 6.2 allows us to interchange the roles of zero and one in the data without affecting the value of (6.55) for any partition. The transformation to do this is

$$x_i' = Hx + b, \quad H = -I, \quad b = (1, \ldots, 1)^T. \tag{6.56}$$

Incidentally, using $H = I$ and b as above we can see that one could code and work with one and two instead of zero and one.

Since the median vectors no longer occur explicitly in the objective function (6.55), the minimal distance method cannot be used with this way of formulating the objective function. In compensation, the exchange method which is in any case better, becomes particularly simple. As always, we used to test for $i \in C_p$, with $m_p > 1$ whether there exists a $j \neq p$ for which

$$e(C_p \setminus \{i\}) + e(C_j \cup \{i\}) < e(C_p) + e(C_j). \tag{6.57}$$

and then we need to select the best such j. When $x_i$ is removed from $C_p$, the number $m_p$ is reduced by one, and $a_{pk}$ is reduced by one whenever $x_{ik} = 1$. We therefore have

$$e(C_p \setminus \{i\}) = \sum_{k=1}^{s} \min (a_{pk} - x_{ik}, m_p - 1 - a_{pk} + x_{ik}). \tag{6.58}$$

Similarly,

$$e(C_j \cup \{i\}) = \sum_{k=1}^{s} \min (a_{jk} + x_{ik}, m_j + 1 - a_{jk} - x_{ik}). \tag{6.59}$$

Just like $e(C_p)$ and $e(C_j)$, expressions (6.58) and (6.59) can be calculated very easily if the $a_{jk}$ ($j=1, \ldots, n$, $k=1, \ldots, s$) are computed once for the starting partition and then updated whenever a successful exchange in the sense of (6.57) is made.

## 6.7 The exchange method for ordinal data

A vector with s components is called an ordinal vector if each of its components only takes on the values $1, \ldots, t_k$ ($k=1, \ldots, s$), and if these values signify a ranking. For example, on a questionnaire, the possible answers to a question might be 'very good', 'good', 'adequate', 'not so good', and 'useless', and these might be coded as 1 to 5. As we have seen in section 6.1, the use of the median is appropriate for this type of data. For ordinal vectors

$$x_i \in \mathop{\times}_{k=1}^{s} \{1, \ldots, t_k\} \quad (i=1, \ldots, m) \tag{6.60}$$

the $L_1$-criterion can be simplified, though not so much as in the special case of $t_k = 2$ ($k=1, \ldots, s$) dealt with in the previous section.

Again, let us consider one component at a time and suppress the suffix k in $t_k$. As in (6.52) we need to calculate

$$t(C_j) = \min_{y \in \mathbb{R}} \sum_{l=1}^{t} a_\ell |\ell - y| , \tag{6.61}$$

where $a_\ell$ is the number of $x_i$ with $x_i = \ell$ ($\ell = 1, \ldots, t$). Since y must similarly be one of the numbers $1, \ldots, t$, we have

$$t(C_j) = \min_{\ell=1,\ldots,t} f_\ell \tag{6.62}$$

where

$$f_\ell = \sum_{r=1}^{\ell-1} a_r (\ell - r) + \sum_{r=\ell+1}^{t} a_r (r - \ell) . \tag{6.63}$$

The expressions $f_\ell$ can be calculated very simply. Since clearly

$$f_{\ell+1} - f_\ell =: g_\ell = \sum_{r=1}^{\ell} a_r - \sum_{r=\ell+1}^{t} a_r \tag{6.64}$$

and

$$g_{\ell+1} - g_\ell = 2\, a_{\ell+1} \tag{6.65}$$

hold true, the $f_\ell$ and $g_\ell$ can be evaluated recursively using

$$f_1 = \sum_{r=2}^{t} a_r(r-1) = \sum_{r=2}^{t} \sum_{s=t-r+2}^{t} a_s\,, \tag{6.66}$$

$$g_1 = a_1 - \sum_{r=2}^{t} a_r \tag{6.67}$$

and

$$f_{\ell+1} = f_\ell + g_\ell \qquad (\ell=1, \ldots, t-1)\,, \tag{6.68}$$

$$g_{\ell+1} = g_\ell + 2\, a_{\ell+1} \qquad (\ell=1, \ldots, t-2) \tag{6.69}$$

To determine (6.61), the recursion (6.68) and (6.69) need not necessarily be carried out completely. As in section 6.1, it can be seen that

$$f(y) = \sum_{l=1}^{t} a_\ell\, |\ell - y|$$

is a piecewise linear convex function of y, whose slope in the interval $[\ell, \ell+1]$ is $g_\ell$ ($\ell = 1, \ldots, t-1$). Because of the convexity, the computation of $f_{\ell+1}$ need not be carried and from the first index $\ell$ for which $g_\ell$ becomes non-negative. Thus,

$$t(C_j) = f_r \quad \text{with } r = \min_{\ell=1,\ldots,t} \quad (\ell : g_\ell \geqslant 0)\,. \tag{6.70}$$

Now return to the general situation (6.60). Denote by $a_{\ell jk}$ the number of values $\ell$ ($\ell = 1, \ldots, t_k$) of the $k^{th}$ component ($k = 1, \ldots, s$) of all original vectors $x_i$ with $i \in C_j$ ($j = 1, \ldots, n$). For every j and k, indices $r = r(j, k)$ need to be calculated according to (6.70). The value of the objective function for the partition C under consideration, calculated from (6.29), amounts to

$$D_1(C) = \sum_{j=1}^{n} \sum_{k=1}^{s} f_{r(j,k)}\,. \tag{6.71}$$

For ordinal vectors (6.60), the invariance property of the $L_1$-criterion means that each component can be coded as $1 + b, \ldots, t_k + b$ instead of $1, \ldots, t_k$ (where b is a whole number), or as $t_k, t_k - 1, \ldots, 1$, without thereby changing the value of the objective function.

A simple up- and downdating formula similar to (6.58) and (6.59) for binary vectors cannot be given here, since the index $r = r(j, k)$ in (6.70) is not determined explicitly, but rather by the intermediate results of a recursion process.

Unless one wishes to store intermediate results and to distinguish between a large number of special cases, it would seem to be more appropriate to re-calculate the indices $r = r(j, k)$ and values $f_r = f_{r(j,k)}$ each time, instead of trying to update them.

**Problem 6.1**: Is the $L_1$-criterion monotone?

**Problem 6.2**: Show that, for binary vectors $x_i \in \mathbb{B}^s$ $(i = 1, \ldots, m)$, the p-dependent criteria

$$\min_{C \in P(n,M)} \sum_{j=1}^{n} \min_{y_j \in \mathbb{B}^s} \sum_{i \in C_j} \| x_i - y_j \|_p^p$$

are equivalent for $p \geqslant 1$. What does this mean for $p = 1$ and $p = 2$?

**Problem 6.3**: Assume that the components $x_{ik}$ $(k=1, \ldots, s)$ in the object vectors $x_i \in \mathbb{R}^s$ are all different for $i = 1, \ldots, m$. For each i, denote the components in order of size by $\hat{x}_{ik}$, that is

$$\hat{x}_{i,1} < \hat{x}_{i,2} < \ldots < \hat{x}_{is} \quad (i=1, \ldots, m).$$

Now set $y_{ik} = j$ when $x_{ik} = \hat{x}_{ij}$.

The ordinal vectors $y_i \in \{1, \ldots, s\}^s$ with these components do not depend on any scaling $H = \operatorname{diag}(h_1, \ldots, h_s)$ of the $x_i$ provided that $h_k > 0$ $(k = 1, \ldots, s)$. If you then apply the $L_1$-criterion to the $y_i$, you have a criterion which is invariant to scale transformations. What are the advantages and disadvantages of this approach? How can you save the method, if the $x_{ik}$ for a fixed i are not all different?

# 7 Criteria for given or computed distances not involving centres

## 7.1 Derivation of three criteria

In this chapter we assume that the distance between any two objects is given, in other words, that we have a symmetric matrix $T \in \mathbb{R}(m, m)$ with

$$T = (t_{ih}) = (t_{hi}), \quad h=2, \ldots, m, \; i=1, \ldots, h, \quad t_{ii} = 0 . \tag{7.1}$$

This matrix may either be collected directly, though this tends to be problematic for large m, or it may be computed from a given data matrix using a metric or some other means of measuring distances, for example squared Euclidean distance. We do not assume that the triangular inequality $t_{ih} \leqslant t_{ik} + t_{kh}$ must hold.

Case 6 in the introduction is an example of the first type of situation. For every pair of objects (i, h) there was exactly one value $t_{ih} = k$ with $k \in \{1, \ldots, (m(m-1)/2\}$. The second situation arises when, for the distance measure d, the problem of determining centres via

$$\min_{y} \sum_{i \in C_j} d(x_i, y) \tag{7.2}$$

cannot in practice be solved explicitly, and/or where, for the centres found, no simple up- and downdating formulae for the exchange of an object between clusters can be specified. For otherwise, one of the objective functions from the first five chapters could be used.

The second situation arises, for example, when the distance between x and $y \in \mathbb{R}^s$ is measured by the $L_p$ distance $\| x-y \|_p$ with $p \neq 1, \infty$; or for $x, y \geqslant 0$ if the Jaccard metric [B51] is used:

$$J(x,y) = 1 - \frac{\sum_{k=1}^{s} \min(x_k, y_k)}{\sum_{k=1}^{s} \max(x_k, y_k)} \tag{7.3}$$

$$= \frac{2 \| x - y \|_1}{\| x \|_1 + \| y \|_1 + \| x - y \|_1}, \quad \text{for } x \neq 0 \text{ or } y \neq 0,$$

$$J(0,0) = 0 .$$

The Jaccard metric should be used for binary vectors when the occurrences of the value one, for example, property present, are to be given greater weight than when binary vectors agree in a component with value zero, for example, property absent. Note that the invariance property of the $L_1$-metric for scale transformations with $h_k = \pm 1$ also applies to the Jaccard metric, though not the invariance relative to translations. If the Jaccard metric is to be used for ordinal vectors, the s components should all vary between 0 and $t_k - 1$.

As in case 5, the second situation also arises when there are q different sorts of variables. Here, given vectors x and y are split into q partial vectors $x^{(r)}$ and $y^{(r)}$ $(r=1, \ldots, q)$, each of which contains only one type of variable. For measuring distances between partial vectors, $d_r$ (which depends on the type of variable) is used, and the distance between the given vectors is defined as

$$d(x,y) = \sum_{r=1}^{q} p_r\, d_r\,(x^{(r)}, y^{(r)}), \quad p_r > 0, \quad \sum_{r=1}^{q} p_r = 1. \tag{7.4}$$

If the $d_r$ $(r=1, \ldots, q)$ are all metrics for the partial vectors, then so is d, as can be easily verified. As weights $p_r$ one can, for example, use the relative frequency of the $r^{th}$ type of variable among the total of s variables.

In both cases, one could argue against the introduction of additional criteria (yet to be specified) which make use of the values (7.1) only, by pointing out that there is really no need to determine a centre y as in (7.2), since one of the given objects k could be defined as such. This is indeed the case. Let an object k with

$$\sum_{h \in C_j} t_{kh} = \min_{i \in C_j} \sum_{h \in C_j} t_{ih}, \quad k \in C_j \tag{7.5}$$

be called an object median of the cluster $C_j$.

Such an approach makes sense. In practical problems one often represents a class not by a specially defined and calculated centre y, but by an object q in the class with

$$d(x_q, y) = \min_{i \in C_j} d(x_i, y), \; y \in C_j\,. \tag{7.6}$$

Of course, the index q need not coincide with k above if the distance matrix T was calculated using the distance measure d.

There are a number of arguments against choosing an object median (7.5) and specifying criteria based on it. Firstly, the object median need not be unique, in contrast to the previously examined problems involving centres (7.2), as is shown by the example

$$T = \begin{pmatrix} 0 & 3 & 5 & 1 \\ 3 & 0 & 2 & 4 \\ 5 & 2 & 0 & 6 \\ 1 & 4 & 6 & 0 \end{pmatrix}.$$

Secondly, the determination of an object median (7.5) is a discrete optimization problem. In the general case, this can only be solved by elaborate enumeration. Thirdly, in contrast to mean and median vectors, there are no up- and downdating formulae for object medians.

Thus, if we refrain from determining centres according to (7.2) because it is impossible or too difficult, and if we refrain from determining object medians according to (7.5), then we need to define objective functions which are based solely on the distance values $t_{ih}$ as in (7.1).

If we take as our measure of compactness of a cluster the sum of all distances between pairs of its members, we have as a first objective function

$$M_1(C) = \sum_{j=1}^{n} \Big( \sum_{i \in C_j} \sum_{\substack{h \in C_j \\ h > i}} t_{ih} \Big) . \tag{7.7}$$

If this plausible measure of compactness is weighted by the reciprocal of the number of elements in the cluster, we obtain (see [A20])

$$M_2(C) = \sum_{j=1}^{n} \Big( \frac{1}{m_j} \sum_{i \in C_j} \sum_{\substack{h \in C_j \\ h > i}} t_{ih} \Big) . \tag{7.8}$$

If we weight m by the reciprocal of the number of distances involved, we obtain

$$M_3(C) = \sum_{j=1}^{n} \Big( \frac{1}{m_j(m_j-1)} \sum_{i \in C_j} \sum_{\substack{h \in C_j \\ h > i}} t_{ih} \Big) . \tag{7.9}$$

When using $M_3$, the partitions C must either be restricted to

$$P^{(2)}(n,M) = \{C : C \in P(n,M), m_j \geqslant 2, j=1, \ldots, n\} \tag{7.10}$$

or a definition

$$m_j = 1 \Rightarrow e(C_j) := 0 \tag{7.11}$$

is required.

In terms of graph theory, $M_1$ can be interpreted as meaning that a complete graph with given edge values is to be split into n complete subgraphs such that the total overall sums of edge values is minimized.

$M_2$ is the same as the variance criterion when $t_{ih} = \| x_i - x_h \|^2$ with $x_i, x_h \in \mathbb{R}^s$. This follows from Theorem 2.3.

Regarding $M_3$, it is worth noting that, where the extra definition (7.11) is used, optimal partitions are frequently those with $m_j = 1$ for $j=1, \ldots, n-1$, and $m_n = m - n + 1$, where $C_n$ is the collection of all those objects for which

$$e(C_n) = \min_{\substack{C \subset M \\ |C| = m-n+1}} \frac{1}{(m-n+1)(m-n)} \sum_{i \in C} \sum_{\substack{h \in C \\ h > i}} t_{ih} \tag{7.12}$$

If, on the other hand, only $C \in P^{(2)}(n,M)$ are admitted, as in (7.10), then optimal partitions are frequently those for which $m_j = 2$ for $j=1, \ldots, n-1$, and $m_n = m - 2(n-1)$. Here, the first $n-1$ clusters each contain two objects which are a small distance apart. We shall see some examples in section 12.3. For these reasons, the objective function $M_3$ must be treated with some caution.

The monotonicity of the objective functions $M_1$, $M_2$, and $M_3$ (the latter at least with (7.11)) is a trivial matter.

## 7.2 Exchange method

There cannot be a minimal distance method of the usual type when an objective function that does not involve centres is used, and specifically, there is none for $M_1$, $M_2$, or $M_3$. The only conceivable variations would work with object medians (7.5) instead of centres. We shall therefore concentrate on the exchange method.

Using $M_1$, we have, for $i \in C_p$ and $m_p > 1$,

$$e(C_p \setminus \{i\}) = e(C_p) - \sum_{h \in C_p} t_{ih} \tag{7.13}$$

and

$$e(C_j \cup \{i\}) = e(C_j) + \sum_{h \in C_j} t_{ih}, \quad j \neq p, \tag{7.14}$$

and the exchange condition (6.75) is reduced to

$$\sum_{h \in C_j} t_{ih} < \sum_{h \in C_p} t_{ih}. \tag{7.15}$$

Using $M_2$, we have

$$e(C_p \setminus \{i\}) = \frac{m_p e(C_p) - \sum_{h \in C_p} t_{ih}}{m_p - 1}. \tag{7.16}$$

$$e(C_j \cup \{i\}) = \frac{m_j e(C_j) + \sum_{h \in C_j} t_{ih}}{m_j + 1}, \tag{7.17}$$

and the exchange condition becomes

$$\frac{\sum_{h \in C_j} t_{ih} - e(C_j)}{m_j + 1} < \frac{\sum_{h \in C_p} t_{ih} - e(C_p)}{m_p - 1}. \tag{7.18}$$

In the same way, the exchange condition for $M_3$ with auxiliary condition (7.10) is

$$\frac{\sum_{h \in C_j} t_{ih} - 2\, m_j\, e(C_j)}{m_j\,(m_j + 1)} < \frac{\sum_{h \in C_p} t_{ih} - 2(m_p - 1)\, e(C_p)}{(m_p - 1)\,(m_p - 2)} . \tag{7.19}$$

Note that (7.10) implies $m_p > 2$.

As always, we are looking for that index j for which the left hand side of (7.15), (7.18) or (7.19) is minimized, provided there is such a j satisfying the inequality.

The exchange formulas for the variance criterion (2.51) required O(s) operations on the left hand side; those for the determinant criterion (3.49) required $O(s^2)$. Here, we require $O(m_j)$ operations, a number which varies from one cluster to another. The storage requirement here is also very much larger, using $m(m-1)/2$ locations for T, compared with the $(m+n)s$ locations used under the variance criterion for storing the data matrix and mean vectors. For large m, say, $m \approx 1000$, the objective functions $M_1$, $M_2$, and $M_3$ present problems of storage and computation time, even on fast computers.

**Problem 7.1:** Let $d: \mathbb{R}^s \times \mathbb{R}^s \to \mathbb{R}$ be a metric, and $h > 0$. Show that the function defined by

$$d_h(x, y) := \frac{d(x, y)}{h + d(x, y)}$$

is also a metric.

**Problem 7.2:** Again, let d be some metric; $x, y, a \in \mathbb{R}^s$, and let a be fixed. Show that the function defined by

$$d_a(x, y) := \frac{d(x, y)}{d(x, a) + d(y, a) + d(x, y)} \quad (x \neq a \;\text{ or }\; y \neq a),\; d_a(a, a) := 0$$

is also a metric. What happens when $a = 0$, when d is the $L_1$ metric, and when there are binary and ordinal vectors?

**Problem 7.3:** For $x_i \in \mathbb{R}^s$, let $x_i \geqslant 0$ $(i=1, \ldots, m)$, and let $d_0^{(1)}$ be that metric which is obtained by substituting $a = 0$ and the $L_1$ metric for d in Problem 7.2. Show that the function

$$f(z) = \sum_{i=1}^{m} d_0^{(1)}(x_i, z)$$

is minimized by some $z \in \mathbb{R}^s$ with $z_k = x_{i_k,k}$ $(k=1, \ldots, s,\; i_k \in \{1, \ldots, m\})$. (See also [B57] and [B58] for a numerical method.)

# 8 Clusterwise linear regression

## 8.1 Derivation of an objective function

The model most frequently used to describe the effect of independent variables $a_1, \ldots, a_s$ on a dependent variable $b = b(a_1, \ldots, a_s)$ is the simplest one, namely

$$b = x_1 a_1 + \ldots + x_s a_s \tag{8.1}$$

with linear parameters $x_1, \ldots, x_s$.

For this, one starts with a matrix $A \in \mathbb{R}(m, s)$ and a vector $b \in \mathbb{R}^m$ of $m \gg s$ observations $((a_{ik}, k=1, \ldots, s), b_i)$ $(i=1, \ldots, m)$ and determines the required vector of parameters $x \in \mathbb{R}^s$ in such a way that the m equations

$$b_i = a_{i1} x_1 + \ldots + a_{is} x_s \quad (i=1, \ldots, m) \tag{8.2}$$

are satisfied as well as possible. However, since such an over-determined system of linear equations is not, in general, exactly solvable, x is determined in such a way that the equations can, in some sense, be solved approximately. With the usual requirement

$$\min_{x \in \mathbb{R}^s} \| Ax - b \|_p , \quad 1 \leqslant p \leqslant \infty , \tag{8.3}$$

the $L_p$-norm of the vector of residuals is to be minimized.

For $p=2$, the term 'regression analysis' is used. Where the aim is merely a description of the data, one can also speak of discrete linear $L_2$- (or $L_p$-) approximation. For this, equivalently, the function

$$F(x) = \| Ax - b \|^2 = (Ax - b)^T (Ax - b) \tag{8.4}$$

needs to be minimized. Since F is strictly convex,

$$\frac{\partial F}{\partial x} = 0 \tag{8.5}$$

provides a necessary and sufficient condition for the existence of a minimum, namely

$$A^T Ax = A^T b. \tag{8.6}$$

These so-called normal equations are uniquely solvable for $m \geqslant s$ if an only if the rank of A equals s. For then the inverse of $A^T A$ exists, and

$$x = (A^T A)^{-1} A^T b. \tag{8.7}$$

For $p \neq 2$, there is no exact formula for the solution. However, the following considerations are also possible for $p \neq 2$.

The use of $1 \leqslant p < 2$, and particularly of $p = 1$, is important where the influence of outliers among the observations on the vector of parameters is to be reduced. This corresponds to the robustness of the median, compared with the sensitivity of the mean to outliers. However, in the following we shall only look at the case of $p = 2$.

If in the task (8.4), or more generally in (8.3), the number of observations m is much greater than the number of explanatory variables s, it may make sense in the interest of a better description to divide the observations $((a_{ik}, k=1, \ldots, s), b_i)$, $i \in M = \{1, \ldots, m\}$ into n clusters $C_j$ $(j=1, \ldots, n)$. Through the influence of omitted or not measurable variables there might even be a structure in the mass of observations.

This suggests that we should seek a partition of length n and corresponding sets of parameters $x^{(j)} \in \mathbb{R}^s$, $j=1, \ldots, n$, such that the sum of the error sums of squares over all clusters, is minimized. If we set

$$(A^{(j)}, b^{(j)}) = ((a_{ik}, k=1, \ldots, s), b_i), i \in C_j, \tag{8.8}$$

this implies the objective function

$$R_2(C) = \sum_{j=1}^{n} \min_{x^{(j)} \in \mathbb{R}^s} \| A^{(j)} x^{(j)} - b^{(j)} \|^2 . \tag{8.9}$$

(see [B47, B48, B49]). To guarantee the existence of the solutions $x^{(j)}$, we require rank $A^{(j)} = s$. A necessary condition for this is $m_j \geqslant s$, which in turn implies $m \geqslant ns$.

If we denote the partitions which are feasible here by

$$P'(n,M) = \{C : C \in P(n,M), \text{rank } A^{(j)} = s, j = 1, \ldots, n \tag{8.10}$$

then we can refer to a partition $C_0$ as optimal when

$$R_2(C_0) = \min_{C \in P'(n,M)} R_2(C). \tag{8.11}$$

In this context, the sets of coefficients $x^{(j)}$ take on the role of the cluster centres. They are the coefficients of n different hyperplanes of the form (8.1).

Let us illustrate this with a rather specialized example. Let points $(t_i, y_i) \in \mathbb{R}^2$, $i=1, \ldots, m$, with $t_1 < t_2 < \ldots < t_m$ be given. At the top of Fig. 8.1 a set of such points is shown together with the straight line of best fit $y = ct + d$. Its coefficients $x = (c, d)^T$ are, in the terminology we have used, a unique solution of the minimization problem (8.3) for $p = 2$, where we equate

$$a_{i1} = t_i, \quad a_{i2} = 1, \quad b_i = y_i, \quad i=1, \ldots, m. \tag{8.12}$$

Clearly, the points drawn in Fig. 8.1 do not have the property that a single straight line constitutes a meaningful description. Lower down, Fig. 8.1 contains the optimal partitions for $n = 2, 3, 4$ which we have required to be contiguous, and which have in this case been found by enumeration, together with the corresponding partial lines of best fit $y = c_j t + d_j$ $(j = 1, \ldots, 4)$.

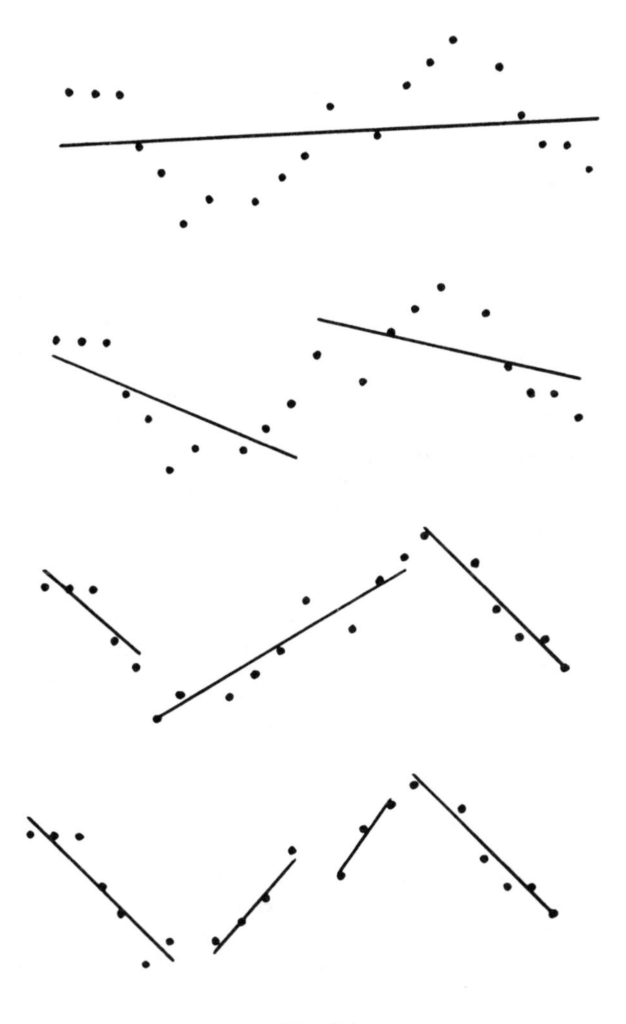

Fig. 8.1

If the $t_i$ are interpreted as points in time, then with n = 3 and/or n = 4, we have obtained a sensible division into time intervals, in which the trend runs in different directions. In the context of a cluster analysis one could speak in terms of finding so-called breaks in the structure.

## 8.2 Minimal distance method

It is of course possible to develop a minimal distance method for the mixed continuous and combinatorial optimization problem (8.11). Starting with a feasible initial partition $C \in P'(n,M)$, centres $x^{(j)}$ can be defined by solving the sub-problems

$$\min_{x^{(j)} \in \mathbb{R}^s} \quad \| A^{(j)} x^{(j)} - b^{(j)} \|^2, \quad j=1,\ldots,n \tag{8.13}$$

A new partition is then detained by reassigning the observations according to the modified minimal distance rule

$$i \in C_j \Longleftrightarrow \sum_{k=1}^{s} (a_{ik} x_k^{(j)} - b_i)^2 = \min_{\ell=1,\ldots,n} \sum_{k=1}^{s} (a_{ik} x_k^{(\ell)} - b_i)^2 . \tag{8.14}$$

Now the centres are re-calculated etc., all on the assumption that the partitions remain feasible. If that is the case, it is possible to demonstrate the contraction property of the minimal distance method, as we have done previously. Unfortunately, however, the method frequently generates classes with $m_j < s$, and this violates a necessary condition for $C \in P'(n,M)$. Therefore we shall once again prefer the exchange method.

The computational effort during each iteration of the minimal n regression problems, each with $m_j$ observations, and of finding m reassignments according to (8.14). Empirical calculations suggest that the overall computational effort when using the exchange method with suitable updating techniques is in general not greater.

## 8.3 Up-and downdating formulae for the solution of regression problems when observations are added are removed; exchange method

With the objective function studied here, namely (8.11) with (8.9) and (8.10), the measure of compactness $e(C_j)$ for a class $C_j$ is given by

$$e(C_j) = \min_{x^{(j)} \in \mathbb{R}^s} \quad \| A^{(j)} x^{(j)} - b^{(j)} \|^2 . \tag{8.15}$$

For each attempted exchange of an $i \in C_p$ with $m_p > s$, n regression problems for $C_p \setminus \{i\}$ and $C_j \cup \{i\}$ $(j=1,\ldots,n, j \neq p)$ need to be solved. The formulation and solution of the corresponding normal equations is very laborious. Consequently, we need to look for suitable up- and downdating formulae, as we did with the previous objective functions. How can regression problems, in which an observation has been removed or added, be solved most simply, using the solution of the original problem? In this section we shall point to a theoretically possible solution (see [B47]), and in the next section and in section 10.5 we shall sketch the practical implementation which uses a so-called QR-decomposition.

Again, we shall simplify the notation for the time being by omitting the cluster indices p and j in $A^{(p)}$, $A^{(j)}$, $b^{(p)}$, and $b^{(j)}$. For the $i^{th}$ observation we shall write

$$(f^T, d) := ((a_{ik}, k=1, \ldots, s), b_i), \quad f \in \mathbb{R}^s, d \in \mathbb{R}. \tag{8.16}$$

When adding the observation $(f^T, d)$ to $(A, b)$, we want to use the solution of

$$\min_{x \in \mathbb{R}^s} \| Ax - b \|^2$$

which is

$$x = (A^T A)^{-1} A^T b \tag{8.17}$$

to calculate the solution $y \in \mathbb{R}^s$ of the problem

$$\min_{y \in \mathbb{R}^s} \left\| \begin{pmatrix} A \\ f^T \end{pmatrix} y - \begin{pmatrix} b \\ d \end{pmatrix} \right\|^2 . \tag{8.18}$$

The solution of (8.18) is given by

$$\begin{aligned} y &= \left( (A^T, f) \begin{pmatrix} A \\ f^T \end{pmatrix} \right)^{-1} \left( (A^T, f) \begin{pmatrix} b \\ d \end{pmatrix} \right) \\ &= (A^T A + f f^T)^{-1} (A^T b + df) \ . \end{aligned} \tag{8.19}$$

If we assume $(A^T A)^{-1}$ to be known, we can use Theorem 3.3, part (b) to calculate $(A^T A + ff^T)^{-1}$, and the required solution y of (8.19) becomes

$$y = z - \frac{g f^T z}{1 + f^T g} \tag{8.20}$$

where

$$g = (A^T A)^{-1} f, \quad z = x + dg \ . \tag{8.21}$$

By hypothesis, we had rank (A) = s. $A^T A$ is therefore positive definite, and $f^T g \geqslant 0$. Since $A^T A$ is positive definite, so is $A^T A + ff^T$. The updating is therefore always possible.

If, on the other hand, a row $f^T$ is removed from A, and the corresponding component d is removed from b, the normal equations (8.6) will change to

$$(A^TA - ff^T)y = A^Tb - df. \tag{8.22}$$

Using Theorem 3.3, part (b), and the symmetry of $A^TA$, we obtain

$$(A^TA - ff^T)^{-1} = (A^TA)^{-1} + \frac{1}{1 - f^T(A^TA)^{-1}f} gg^T \tag{8.23}$$

with

$$g = (A^TA)^{-1}f. \tag{8.24}$$

Thus, $(A^TA - ff^T)^{-1}$ exists and is positive definite whenever

$$0 \leqslant f^T(A^TA)^{-1}f < 1. \tag{8.25}$$

Assuming that (8.25) holds, and using (8.23) and (8.24), the solution y of (8.22) can be expressed as

$$y = z + \frac{gf^Tz}{1 - f^Tg}, \quad z = x - dg. \tag{8.26}$$

Thus, the feasibility of a partition C is retained after the exchange if and only if condition (8.25) is satisfied for $i \in C_p$, $A = A^{(p)}$, and f is defined in (8.16).

## 8.4 Numerical methods

Since the inverses of $(A^TA \pm ff^T)$ are used again explicitly after the up- or downdating in (8.21) or (8.24), the method outlined above effectively requires that the normal equations (8.6) are also solved by calculating the inverse of $A^{(j)T}A^{(j)}$ for each cluster, and that $x^{(j)}$ is obtained using (8.7).

However, as is well known, the number of computations involved in solving a system of linear equations by first calculating the inverse of the matrix of coefficients exceeds that required for the elimination method by a factor of s. It is therefore not used in practice. Instead, a Cholesky decomposition [B54] of the matrices $A^TA$ into

$$A^TA = LL^T \tag{8.27}$$

is performed, where L is a lower triangular matrix. The normal equations (8.17) are solved in two steps by

$$Lz = A^Tb, \quad L^Tx = z \tag{8.28}$$

and g in (8.21) and (8.24) is found by

$$Lz = f, \quad L^Tg = z. \tag{8.29}$$

With this procedure it is not the inverses but the Cholesky decompositions of $(A^TA \pm ff^T)$ that are to be up- and downdated and stored. Although efficient algorithms for this are available [B21, B22] which are also used in the implementation of the criterion of adaptive distances, we suggest a different approach for regression analysis, one which has become widely used and is in general numerically superior.

Instead of constructing and solving the normal equations, we first perform a QR-decomposition of $A \in \mathbb{R}(m, s)$, with

$$A = QR \tag{8.30}$$

where $Q \in \mathbb{R}(m, s)$ is a matrix with orthogonal or orthonormal columns and R is an upper triangular matrix. The normal equations (8.6) then become

$$A^T Ax = (QR)^T QRx = R^T Rx = R^T Q^T b\,. \tag{8.31}$$

The pre-condition rank (A) = s implies that the matrix R is non-singular. Hence, (8.31) can be solved via

$$Rx = Q^T b \tag{8.32}$$

through backward substitution.

The QR-decomposition (8.30) corresponds to orthogonalization or orthonormalization with the aid of the Gram-Schmidt orthogonalization method of linear algebra. Numerically, it is performed using a modified Gram-Schmidt method (in which the elements of R are calculated in a different order) or using a Householder or Givens transformation [B29]. Each of these numerical methods has its advantages and disadvantages. All are recognized to be numerically preferable to the approach using normal equations [B54].

In our context, an up- and downdating of the QR-decomposition of A becomes necessary whenever a row is added or deleted. These methods have also been developed [B11], but we shall not describe them in detail either here or in the implementation section of this book, since they can by now be found in standard textbooks on numerical methods [B54].

## 8.5 Use of a discrete $L_p$-approximation

If an $L_p$-norm is used in (8.9), the objective function to be minimized for $C \in P'(n,M)$ becomes

$$R_p(C) = \sum_{j=1}^{n} \min_{x^{(j)} \in \mathbb{R}^s} \| A^{(j)} x^{(j)} - b^{(j)} \|_p^p\,, \quad p \geqslant 1\,. \tag{8.33}$$

For $p \neq 2$, the solution $x^{(j)}$ cannot be specified explicitly. For p in the range $1 < p < \infty$, an iterative method is necessary [B33, B38, B56, B57]. For $p = 1$ and $p = \infty$, the problems can be transformed into special linear programs whose solutions need no longer by unique [B3]. The reason is that the $L_p$-norm for $1 < p < \infty$ is strictly convex, whereas it is merely convex for $p = 1, \infty$.

Except for $p = 2$, the only hope of finding up- and downdating formulae which apply when an observation is added or removed, is for $p = 1$ and $p = \infty$. Here, methods are required which efficiently update the solutions of linear programs offer the removal or addition of rows or columns, depending on whether a primal [B6] or dual [B1] method is employed [B3, B36].

For $p = 1$, these methods are developed and implemented in [B36]. The basis for an implementation of the exchange method therefore exists. The criterion $R_1 = R_1$ (C) is of particular importance, and preferable to $R_2$, where outliers among the observations are suspected.

**Problem 8.1:** Is the criterion (8.9) or (8.33) monotone for $p \neq 2$? Consider especially the case $p = 1$.

**Problem 8.2:** Develop updating formulae for the two coefficients in a least squares fit using a straight line. Can you derive a heuristic or an exact method for determining solutions such as those shown in Fig. 8.1?

# Part II

# Implementation of FORTRAN subroutines

# 9 Implementations for the minimum variance and determinant criteria, and for the criterion of adaptive distances

## 9.1 Basic notes on the implementation, and some auxiliary routines

Standard FORTRAN is the language used in the implementation of the approximation methods associated with the various criteria. The use of the simplest language elements possible enhances the portability of the routines. They will run on any machine with a FORTRAN compiler that includes FORTRAN IV as a subset. Occasionally it might be appropriate to adapt some constants, for example BIG (big number) and EPS (precision test). Usually, such changes will not be necessary, since the values used are not set to match the greater numeric range and accuracy of the TR440 used for testing (11 decimal places), but are appropriate for the range and accuracy of IBM machines.

Clarity and legibility were sought when writing the programs. All but one of the programs incorporate versions of the exchange method which share a common structure. Once the relevant subroutine for the variance criterion (TRWEXM) has been studied closely, the other versions are easy to understand. The names of subroutine parameters and other variables match the mathematical notation employed in the preceding eight chapters as closely as possible. Thus, M always denotes the number m of objects, S the number of s variables, N the number n of clusters, and M0 the prescribed minimum number $m_0$ of cluster members. All parameters are explained in the text when they first occur, and they retain their meaning unless it is changed explicitly. Consequently, it was possible to omit comment cards in the programs entirely.

To allow the use of variable dimensions with two- and three-dimensional arrays which are passed as parameters into subroutines (higher dimensions do not occur), the subroutine parameter list must include both the dimensions (for example M and N) with which the array is used in the body of the subroutine, and the first, or first and second, maximum dimensions with which the array is originally declared in the calling program. The latter would be denoted by MDX and NDX respectively. We shall explain this with reference to subroutine TRAFOR in Table 9.1, which takes a given data matrix ((X(I,K), K=1, S), I=1, M), performs the standardization (2.31) to (2.33), and overwrites X with X′. In

TRAFOR, the dimension of X is declared to be X(MDX,S). If a main program is to call TRAFOR for data matrices with $1 \leqslant M \leqslant 200$ and $1 \leqslant S \leqslant 10$, then the dimension of X is declared as X(200,10) in the main program, and TRAFOR is called with

CALL TRAFOR (X, 200, M, S, IFLAG)

where $1 \leqslant M \leqslant 200$. The subroutine does not check whether $M \leqslant MDX$. If he wishes, the user must do this himself before calling the routine. The storage requirement of the various subroutines – in this case ms – can easily be deduced from the DIMENSION statement, and is therefore not mentioned explicitly.

*Table 9.1*

```
      SUBROUTINE TRAFOR (X,MDX,M,S,IFLAG)
      INTEGER   S
      DIMENSION X(MDX,S)
      IFLAG=0
      DO 5 K=1,S
           H=0.
           DO 1 I=1,M
                H=H+X(I,K)
1          CONTINUE
           XBAR=H/M
           H=0.
           DO 2 I=1,M
                F=X(I,K)-XBAR
                H=H+F*F
2          CONTINUE
           IF(H.GT.0.) GOTO 3
           IFLAG=8
           GOTO 6
3          H=1./SQRT(H)
           DO 4 I=1,M
                X(I,K)=H*(X(I,K)-XBAR)
4          CONTINUE
5     CONTINUE
6     RETURN
      END
```

Errors detected inside the subroutines are trapped via the parameter IFLAG. If IFLAG = 0 after a call, no error has been detected. Positive values of IFLAG have the same meaning for all subroutines. They will be explained when they first occur. Thus, the error indicator IFLAG = 8 in TRAFOR means that the result of calculating $s_k$ according to (2.32) is either zero, or that rounding errors have caused $s_k^2 < 0$ for some $k=1, \ldots, s$, so that the square root could not be calculated. When IFLAG is non-zero, the output parameters are not in general set to the right values, and can therefore not be used further.

Subroutine MEANS in Table 9.2 serves as the subprogram for computing the mean vectors for a given partition. It is used whenever the cluster centres are mean vectors. (Table 13.1 contains a summary of which programs and subroutines call which other subroutines.) Besides MDX, M, S, M0, NDX, and N, MEANS expects a data matrix ((X(I, K), K=1, S), I=1, M) and a vector of integers (Z(I), I=1, M) which characterizes the partition to be dealt with. Z(I) must be set equal to J if the $I^{th}$ object is to belong to the $J^{th}$ cluster. If the partition thus characterized is feasible, that is, if $1 \leqslant Z(I) \leqslant N$ (I=1, M) and if there are for each J=1, N at least M0 indices with Z(I) = J, then the numbers

*Table 9.2*

```
      SUBROUTINE MEANS (X,MDX,M,S,Z,MO,MJ,XBAR,NDX,N,IFLAG)
      INTEGER    Z(M),MJ(N),S
      DIMENSION X(MDX,S),XBAR(NDX,S)
      IFLAG=0
      DO 2 J=1,N
           MJ(J)=0
           DO 1 K=1,S
                XBAR(J,K)=0.
    1      CONTINUE
    2 CONTINUE
      DO 5 I=1,M
           J=Z(I)
           IF(J.GE.1.AND.J.LE.N) GOTO 3
           IFLAG=1
           GOTO 9
    3      MJ(J)=MJ(J)+1
           DO 4 K=1,S
                XBAR(J,K)=XBAR(J,K)+X(I,K)
    4      CONTINUE
    5 CONTINUE
      DO 8 J=1,N
           L=MJ(J)
           IF(L.GE.MO) GOTO 6
           IFLAG=2
           GOTO 9
    6      F=1./L
           DO 7 K=1,S
                XBAR(J,K)=XBAR(J,K)*F
    7      CONTINUE
    8 CONTINUE
    9 RETURN
      END
```

*Table 9.3*

```
      SUBROUTINE TRACES (X,MDX,M,S,Z,XBAR,NDX,N,E,D,IFLAG)
      INTEGER    Z(M),S
      DIMENSION X(MDX,S),XBAR(NDX,S),E(N)
      D=0.
      DO 1 J=1,N
           E(J)=0.
    1 CONTINUE
      IF(IFLAG.NE.0) GOTO 4
      DO 3 I=1,M
           J=Z(I)
           H=0.
           DO 2 K=1,S
                F=XBAR(J,K)-X(I,K)
                H=H+F*F
    2      CONTINUE
           E(J)=E(J)+H
           D=D+H
    3 CONTINUE
    4 RETURN
      END
```

$m_j$ of cluster members are computed and returned in (MJ(J), J=1, N), and the mean vectors $\bar{x}_j$ in ((XBAR(J, K), K=1, S), J=1, N). If Z(I) < 1 or Z(I) > N for some I, returns the error IFLAG = 1. If MJ(J) < M0 for some J, MEANS returns IFLAG = 2 without computing the desired results. M0 ⩾ 1 is assumed.

The subroutine TRACES in Table 9.3 is called following a call to MEANS, and calculates the measures of compactness $e_j$ according to (2.6), placing the result in (E(J), J = 1, N). In addition, the sum D of the $e_j$ is calculated for the given partition. This is the value of the variance criterion. The parameter IFLAG is included for formal consistency and has no significance in this instance.

## 9.2 TRWMDM: Minimal distance method for the variance criterion

Starting with an initial partition given by (Z(I), I = 1, M) whose feasibility has been checked by MEANS, TRWMDM is called to perform the minimal distance method as described in Section 2.7, but with a modification which we shall explain below.

The only new parameter is IT. On exit, IT contains the number of iterations performed. Since at most five to ten iterations are needed (see also Section 11.4), their number is internally limited to ITMAX = 15. The user should therefore observe the value of IT.

If more than 15 iterations are to be performed, the relevant test can either be changed, or the end partition (Z(I), I = 1, M) resulting from a first call to TRWMDM may be used as input to a second call.

The important modification compound with the convergence test described in step 3, namely

$$D(C^{(t+1)}) < D(C^{(t)}) \tag{9.1}$$

and

$$D(C^{(t+1)}) \leqslant D(C^{(t)}), \quad t \leqslant \text{ITMAX}, \tag{9.2}$$

consists of using instead the test

$$D(C^{(t+1)}) \leqslant R\, D(C^{(t)}), \quad 0 < R < 1, \tag{9.3}$$

where t ⩽ ITMAX = 15. In TRWMDM, R = .999. Although this change can in principle lead to different convergence behaviour, it has the effect of stopping any cyclic divergence which may result from rounding errors, before the maximum number of iterations is reached. In the implementation of the exchange method, we shall have to discuss this eventuality more fully.

For t = 0, the comparison (9.3) is made with

$$D(C^{(0)}) = \text{DMAX} = \text{BIG} \tag{9.4}$$

where BIG is a very large number on the computer used. Here, as in all programs in which BIG plays a similar role, we have set BIG = 1.E50. This may need to be changed. If it should happen that for some starting partition

$$D(C^{(1)}) \geqslant \text{BIG} \tag{9.5}$$

and that this number is still within the range that the computer can handle, then TRWMDM will stop with IT = 1. The same applies to all later subroutines in which BIG is defined and used in this way.

If a minimal distance iteration produces a class with fewer than M0 elements, MEANS will recognize this fact, and TRWMDM will return an error IFLAG = 2. In general, M0 would be specified as 1, unless one wished to fulfil auxiliary conditions as in (5.50). After a return with IFLAG = 0, the parameters Z, MJ, XBAR, E, D, and IT hold the values for the end partition found.

*Table 9.4*

```
      SUBROUTINE TRWMDM (X,MDX,M,S,Z,MO,MJ,XBAR,NDX,N,E,D,IT,IFLAG)
      INTEGER    Z(M),MJ(N),S
      DIMENSION X(MDX,S),XBAR(NDX,S),E(N)
      BIG=1.E50
      R=.999
      DMAX=BIG
      IT=0
    1 IT=IT+1
      IF(IT.GT.15) GOTO 5
      CALL MEANS (X,MDX,M,S,Z,MO,MJ,XBAR,NDX,N,IFLAG)
      CALL TRACES (X,MDX,M,S,Z,XBAR,NDX,N,E,D,IFLAG)
      IF(IFLAG.NE.0) GOTO 5
      IF(N.LE.1) GOTO 5
      IF(D.GE.DMAX) GOTO 5
      DMAX=D*R
      DO 4 I=1,M
            F=BIG
            DO 3 J=1,N
                  H=0.
                  DO 2 K=1,S
                          T=XBAR(J,K)-X(I,K)
                          H=H+T*T
    2             CONTINUE
                  IF(H.GE.F) GOTO 3
                  F=H
                  L=J
    3       CONTINUE
            Z(I)=L
    4 CONTINUE
      GOTO 1
    5 RETURN
      END
```

If we calculate the computation time by counting all arithmetic operations as equal units and by ignoring the effect of index operations and subroutine calls which are more difficult to estimate, then approximately 4ms + ns + 5m + 3mns + mn units are needed for each minimal distance iteration. We shall not do these operation counts on a systematic basis, especially as it will not be possible with the exchange method. Instead, we shall provide examples of actual computation times in later sections.

## 9.3 TRWEXM: Exchange method for the variance criterion

Starting from an initial partition given by (Z(I), I=1, M) whose feasibility is tested in MEANS, TRWEXM performs the exchange method for tr **W** as described in Section 2.8, but with one important modification. This regards the test of whether an exchange has been successful. Provided there exist indices j with $f_j > 0$ according to (2.49), the transfer to q according to (2.50) is performed only if

$$\frac{m_q}{m_q+1} \| x_i - \bar{x}_q \|^2 < R \frac{m_p}{m_p-1} \| x_i - \bar{x}_p \|^2, \quad 0<R<1. \tag{9.6}$$

The quantity R plays the same role as in (9.3) but is now used in another inequality. The equivalent test has been similarly modified in almost all programs that use the exchange method, except where integer arithmetic is used. We always set R = .999. This prevents an object being repeatedly moved from one class into another and back again in the next pass, as a result of rounding errors.

We give an example for the TR 440 computer where this would actually happen if we had R = 1. Let m = 11, s = 2, and X given by

$$X^T = \begin{pmatrix} 298 & 299 & 300 & 301 & 302 & 350 & 398 & 399 & 400 & 401 & 402 \\ 202 & 201 & 200 & 199 & 198 & 250 & 202 & 201 & 200 & 199 & 198 \end{pmatrix}.$$

Consider one of the two optimal partitions of length n = 2, namely $C_1 = \{1, \ldots, 6\}$ and $C_2 = \{7, \ldots, 11\}$ with the mean vectors $\bar{x}_1 = (1850/6, 1250/6)^T$ and $\bar{x}_2 = (400, 200)^T$.

For the attempted exchange of i = 6 from $C_1$ to $C_2$ we have

$$f_1 = \frac{m_1}{m_1-1} \| x_6 - \bar{x}_1 \|^2 = \frac{6}{5} \left( \left(350 - \frac{1850}{6}\right)^2 + \left(250 - \frac{1250}{6}\right)^2 \right)$$

and

$$f_2 = \frac{m_2}{m_2+1} \| x_6 - \bar{x}_2 \|^2 = \frac{5}{6} \left( (350-400)^2 + (250-200)^2 \right).$$

In exact arithmetic,

$$f_1 = f_2 = \frac{12500}{3}.$$

However, the TR 440 computer finds $f_2 < f_1$, and so the exchange is performed when R = 1. Subsequently, the mean vectors are updated. No other objects are exchanged. At the next stage of the procedure, the test for moving i = 6 from $C_2$ back to $C_1$ is carried out. In exact arithmetic

$$f_1 = \frac{m_2}{m_2-1} \| x_6 - \bar{x}_2 \|^2, \quad f_2 = \frac{m_1}{m_1+1} \| x_6 - \bar{x}_1 \|^2$$

would again be equal. On the TR 440, however, $f_2 < f_1$, and the exchange would take place. In this special situation the numbers on the computer are such that there would be constant swapping back and forth, and after hundreds of passes the value of the objective function would have changed only in the seventh decimal place, compared with the initial value. By imposing R = .999, this behaviour is prevented.

One must, however, be aware that this modification in the implementation of the exchange method might result in behaviour which differs from that described in section 2.8.

As in TRWMDM, M0 determines the minimum cluster size in TRWEXM. If $m_j < M0$ in a starting partition, there is an exit from TRWEXM via MEANS, with IFLAG = 2. In general, M0 is set to 1. Where there are subsidiary conditions on the form

$$m_j \geqslant m_0 \geqslant 1, \tag{9.7}$$

objects $i \in C_p$ with $m_p = m_0$ are not treated as candidates for an exchange. One should note that with $m_0 > 1$, the exchange method will not work as well as with $m_0 = 1$ [B55].

The quantity BIG is used in TRWEXM to have a comparison value for the first j with $f_j > 0$ at the start of the recursive computation of $f_q$ according to (2.50). Since BIG = 1.E50, the method will fail as described, if one of the $f_j$ with $f_j > 0$ is greater than 1.E50.

With big data matrices and unfortunate combination of numbers, it is possible that rounding errors in the updating of mean vectors and of measures of compactness may lead to an end partition which differs from one which would have been obtained by using exact arithmetic. To prevent this, double precision could be used, although the problem could still occur. With TRWEXM, as well as with all other subroutines that perform the exchange method with floating point arithmetic, it is advisable to call the subroutine a second time, using the end partition found previously as the starting partition. In this way, the mean vectors and measures of compactness would be calculated anew by MEANS and TRACES, and there would be at least one further exchange attempt for each object.

In principle, the number of computer operations needed by TRWEXM could simply be counted in the same units as those used with TRWMDM, and the same could be done for all subsequent subroutines. But since that number also depends on the unknown number of iterations, and on the unknown number of successful exchange attempts, we shall refrain from reproducing the resulting counts.

Note also that TRWMDM and TRWEXM could be reorganised by integrating MEANS and TRACES, so as to allow the rows of the data matrix to be processed sequentially (see [A1, A22]). This makes it possible to hold the data matrix on external, sequentially accessed, storage devices, and hence to process almost arbitrarily large numbers of objects m. However, since realistic cases with $m \approx 5000$ and $s \approx 10$ create no storage problems on the larger computers of today, and since the directly accessible storage is likely to be even greater in future, we have refrained from implementing the method using sequential access. This remark also holds true for the other objective functions which work with data matrices.

*Table 9.5*

```
      SUBROUTINE TRWEXM (X,MDX,M,S,Z,MO,MJ,XBAR,NDX,N,E,D,IT,IFLAG)
      INTEGER   Z(M),MJ(N),S,P,Q
      DIMENSION X(MDX,S),XBAR(NDX,S),E(N)
      BIG=1.E50
      R=.999
      CALL MEANS (X,MDX,M,S,Z,MO,MJ,XBAR,NDX,N,IFLAG)
      CALL TRACES (X,MDX,M,S,Z,XBAR,NDX,N,E,D,IFLAG)
      IF(IFLAG.NE.0) GOTO 8
      IF(N.LE.1) GOTO 8
      IT=0
      IS=0
      I=0
    1 IS=IS+1
      IF(IS.GT.M) GOTO 8
    2 I=I+1
      IF(I.LE.M) GOTO 3
      IT=IT+1
      IF(IT.GT.15) GOTO 8
      I=1
    3 P=Z(I)
      L=MJ(P)
      IF(L.LE.MO) GOTO 1
      V=L
      EQ=BIG
      DO 6 J=1,N
            H=0.
            DO 4 K=1,S
                  T=XBAR(J,K)-X(I,K)
                  H=H+T*T
    4       CONTINUE
            IF(J.NE.P) GOTO 5
            F=V/(V-1.)
            EP=H*F
            GOTO 6
    5       U=MJ(J)
            F=U/(U+1.)
            EJ=H*F
            IF(EJ.GE.EQ) GOTO 6
            EQ=EJ
            Q=J
            W=U
    6 CONTINUE
      IF(EQ.GE.EP*R) GOTO 1
      IS=0
      E(P)=E(P)-EP
      E(Q)=E(Q)+EQ
      D=D-EP+EQ
      FP=1./(V-1.)
      FQ=1./(W+1.)
      DO 7 K=1,S
            H=X(I,K)
            XBAR(P,K)=(V*XBAR(P,K)-H)*FP
            XBAR(Q,K)=(W*XBAR(Q,K)+H)*FQ
    7 CONTINUE
      Z(I)=Q
      MJ(P)=L-1
      MJ(Q)=MJ(Q)+1
      GOTO 2
    8 RETURN
      END
```

## 9.4 DETEXM: Exchange method for the Determinant Criterion

The empirical comparison of TRWMDM and TRWEXM in Section 11.4 will show that the exchange method is superior to the minimal distance method both in terms of computation time and of the values of the objective function for the end partitions found. Although the statement about computation time is not necessarily true for the determinant criterion and the criterion of adaptive distances, we shall refrain from implementing the minimal distance method for these criteria. There are two reasons. Firstly, the method frequently fails when the minimal distance step produces partitions that are not feasible. Secondly, the values of the objective function are generally worse, as is documented in [B35] for the criterion of adaptive distances.

With the determinant criterion, we can implement the exchange method in three ways. Three mathematically equivalent formulations, (3.44), (3.49) and (3.51), are available. (3.44) requires the computation of the determinant of $W(C'')$ which is derived from $W(C)$ by subtraction and addition of a dyadic product. Using (3.49), expressions of the form $x^T A^{-1} x$ with $A = W(C')$ are to be calculated; using (3.51), expressions like $x^T A^{-1} x$ and $x^T A^{-1} y$ with $A = W(C)$ are required. In all three cases, a Cholesky decomposition of the positive definite matrix A is useful. We perform this in the form

$$A = LDL^T \tag{9.8}$$

where L is a lower triangular matrix (different from the one used before) with ones in the main diagonal, and D is a diagonal matrix with positive elements in the diagonal. Compared with the usual Cholesky decomposition, this so-called $LDL^T$ decomposition has the advantage that no roots need to be calculated, and that

$$\det A = \det D = \prod_{k=1}^{s} d_{kk} \tag{9.9}$$

since $\det L = 1$. The expressions needed for the other two variants can also be simply calculated via

$$a = L^{-1}x, \quad x^T A^{-1} x = a^T D^{-1} a \tag{9.10}$$

and

$$a = L^{-1}x, \quad b = L^{-1}y, \quad x^T A^{-1} y = a^T D^{-1} b\,. \tag{9.11}$$

Before we decide on one of the variants, we must ensure that we can update the $LDL^T$ decomposition whenever dyadic products are added to or subtracted from A. This requires that there should be a simple way of calculating $L'$ and $D'$ with the same properties as L and D, for

$$A' = A + \alpha vv^T = L'D'L'^T. \tag{9.12}$$

Here, we must assume that $A'$ is positive definite in order to guarantee the existence of the decomposition. The vector p satisfying

$$Lp = v \tag{9.13}$$

can easily be calculated using forward substitution, and for it, the identity

$$A + \alpha vv^T = L(D + \alpha pp^T)L^T \tag{9.14}$$

holds. Here, the matrix $D + \alpha pp^T$ is of such a special form that one can expect its $LDL^T$ decomposition

$$D + \alpha pp^T = \tilde{L}\tilde{D}\tilde{L}^T \tag{9.15}$$

to be easy to calculate. This is indeed the case [B21]. Once $\tilde{L}$ and $\tilde{D}$ have been obtained (9.15) and (9.14) and (9.12) give

$$A' = L\tilde{L}\tilde{D}\tilde{L}^T L^T. \tag{9.16}$$

Since $L\tilde{L}$ is the product of two lower triangular matrices with ones in the diagonal, it is also of that form. Hence, for $\mathrm{diag}(\tilde{D}) > 0$,

$$L' = L\tilde{L}, \quad D' = \tilde{D} \tag{9.17}$$

is a decomposition satisfying (9.12), as required.

For the numerical computation of $D'$ and $L'$ we use the procedure C1, one of many variations given in [B21] and [B22], without the modification mentioned in [B22]. C1 is the fastest procedure, using $O(s^2)$ multiplications and additions. Numerical stability is assured, provided that the matrix A is 'sufficiently' positive definite, that is, provided that its smallest eigenvalue is sufficiently large relative to the largest one or to a norm of $A'$. If $\alpha < 0$ and $A'$ is nearly singular, the procedure may fail due to rounding errors, in that the diagonal elements of $D'$ become very small or even negative. This cannot be the case if $A'$ is to be positive definite, as we require. We shall later make a suggestion about preventative measures that can be taken in the case of the scatter matrices to be downdated here.

In the usual, easily modified way [B54], subroutine LDLT shown in Table 9.6 calculates the $LDL^T$ decomposition of a positive definite matrix A according to (9.8). ((A(I,J), J=I, S), I=1, S) is the (symmetric) matrix to be split up, (V(I), I=1, S), is work space. The result is stored in ((A(I, J), J=1, S), I=J, S). When the procedure fails owing to rounding errors or because the matrix A was numerically not positive definite (that is, if $d_{kk} <$ EPS for some k=1, ..., s), IFLAG is set to 4 on return from LDLT. Otherwise, IFLAG = 0, and A contains the required $LDL^T$-decomposition. The number of additions and multiplications required as $O(s^3/6)$.

*Table 9.6*

```
      SUBROUTINE LDLT (S,SDX,A,V,EPS,IFLAG)
      INTEGER   S,SDX
      DIMENSION A(SDX,S),V(S)
      IFLAG=0
      DO 6 I=1,S
           DO 5 J=I,S
                H=A(I,J)
                IF(I.EQ.1) GOTO 2
                I1=I-1
                DO 1 K=1,I1
                     H=H-V(K)*A(J,K)*A(I,K)
1               CONTINUE
2               IF(I.NE.J) GOTO 4
                IF(H.GE.EPS) GOTO 3
                IFLAG=4
                GOTO 7
3               V(I)=H
                A(I,I)=H
                GOTO 5
4               A(J,I)=H/V(I)
5          CONTINUE
6     CONTINUE
7     RETURN
      END
```

Subroutine UPDATE in Table 9.7 uses an $LDL^T$ decomposition of A, produced by LDLT and stored in ((A(I, K), K=1, S), I=K, S) as input, together with the given factor ALPHA = $\alpha$ and the given vector (V(I), I = 1, S), as in

*Table 9.7*

```
      SUBROUTINE UPDATE (S,SDX,ALPHA,V,A,DET,EPS,IFLAG)
      INTEGER   S,SDX
      DIMENSION A(SDX,S),V(S)
      IFLAG=0
      DET=1.
      DO 3 J=1,S
           PJ=V(J)
           H=ALPHA*PJ
           DJ=A(J,J)
           DJBAR=DJ+H*PJ
           IF(DJBAR.GE.EPS) GOTO 1
           IFLAG=6
           GOTO 4
1          A(J,J)=DJBAR
           DET=DET*DJBAR
           IF(J.EQ.S) GOTO 3
           BETAJ=H/DJBAR
           ALPHA=DJ*ALPHA/DJBAR
           J1=J+1
           DO 2 L=J1,S
                ALJ=A(L,J)
                V(L)=V(L)-PJ*ALJ
                A(L,J)=ALJ+BETAJ*V(L)
2          CONTINUE
3     CONTINUE
      IF(DET.GE.EPS) GOTO 4
      IFLAG=7
4     RETURN
      END
```

(9.12). UPDATE calculates the $LDL^T$ decomposition of A′ and stores it in ((A(I,K), K=1, S), I=K,S). In addition, the determinant of A′ is calculated and returned in DET. This happens provided that IFLAG = 0. If, during the computation, a $\tilde{d}_{kk}$ < EPS or DET < EPS is encountered, UPDATE returns the error indication IFLAG = 6 or IFLAG = 7, respectively.

If one compares the amount of computing required for the three possible forms of exchange conditions for an $i \in C_p$, one can see that it is of the order of magnitude $O(ns^2)$ in all three cases (3.44), (3.49) and (3.51). In our implementation, we have chosen the form (3.49) in analogy with the variance criterion.

To calculate the squared distance

$$\text{DISTJ} = (x_i - \bar{x}_j)^T \, W(C')^{-1} \, (x_i - \bar{x}_j) \tag{9.18}$$

of a vector (XI(K), K=1, S) from its mean vector (XBAR(J, K), K=1, S) with J = Z(I) in terms of the metric defined by $W(C')^{-1}$, we use the subroutine DISTW shown in Table 9.8. As input, the routine expects the $LDL^T$ decomposition of W(C′) to be supplied in ((A(I, K), K=1, S), I=K, S).

*Table 9.8*

```
      SUBROUTINE DISTW (XBAR,NDX,N,XI,A,SDX,S,J,DISTJ,V)
      INTEGER    SDX,S
      DIMENSION XBAR(NDX,S),XI(S),A(SDX,S),V(S)
      DISTJ=0.
      DO 3 K=1,S
         H=XBAR(J,K)-XI(K)
         IF(K.EQ.1) GOTO 2
         K1=K-1
         DO 1 L=1,K1
            H=H-A(K,L)*V(L)
1        CONTINUE
2        V(K)=H
         DISTJ=DISTJ+(H/A(K,K))*H
3     CONTINUE
      RETURN
      END
```

The subroutine DETEXM shown in Table 9.9 implements the exchange method for the determinant criterion as follows. After MEANS has been called to get the mean vectors, the scatter matrix within the classes, that is, W as defined in (2.19), is determined, and LDLT computes its $LDL^T$ decomposition for the given starting partition (Z(I), I=1, M). Then exchange attempts are made using UPDATE and DISTW. The program structure is identical to that of TRWEXM. The following important modification has been incorporated in DETEXM since it has sometimes proved necessary. Where a successful exchange according to (3.49) is possible, but where rounding errors are such that the mathematically equivalent inequality (3.44) does not hold, the exchange is not carried out, and the method proceeds to the next object. The constant R, discussed in the context of TRWEXM, has the same function as in that routine.

This time it is used on the right hand side of (3.49). Similarly, BIG has the same meaning. Additional output parameters are the scatter matrix ((W(K, L), L=K, S), K=1, S) within classes, and the value DETW of its determinant for the end partition. The arrays WBAR, XI and F are workspace of dimension as indicated.

At the beginning of the DETEXM the three variables EPS1, EPS2 and EPS3 are assigned specific values. EPS1 corresponds to EPS in UPDATE, EPS3 to EPS in LDLT. When the value of the objective function applied to the starting partition is smaller the EPS2, DETEXM returns IFLAG = 9. If, for a given data matrix, there are frequent occurrences of IFLAG = 4, 6, 7 or 9 under a variety of starting partitions, then an adjustment of EPS1, EPS2 and EPS3 might help. This is, of course, machine dependent. In this case, it may also be that ns is too small in comparison with m to make the matrix W 'sufficiently' positive definite in the sense described previously. Perhaps, an increase of M0, which in general ought to equal one, to M0 > 1, may also be helpful. In accordance with the function of this parameter, the exchange method would then no longer be applied in the form described.

*Table 9.9*

```
      SUBROUTINE DETEXM (X,MDX,M,SDX,S,Z,MO,MJ,XBAR,NDX,N,
     *                   DETW,IT,IFLAG,W,WBAR,XI,F)
      INTEGER    Z(M),MJ(N),S,SDX,P,Q
      DIMENSION X(MDX,S),XBAR(NDX,S),F(S),
     *          W(SDX,S),WBAR(SDX,S),XI(S)
      IT=0
      EPS1=1.E-6
      EPS3=1.E-6
      EPS2=1.E-6
      BIG=1.E50
      R=.999
      DETW=0.
      CALL MEANS (X,MDX,M,S,Z,MO,MJ,XBAR,NDX,N,IFLAG)
      IF(IFLAG.NE.0) GOTO 18
      DO 1 K=1,S
      DO 1 L=K,S
    1 W(K,L)=0.
      DO 5 I=1,M
           J=Z(I)
           DO 2 K=1,S
                F(K)=XBAR(J,K)-X(I,K)
    2      CONTINUE
           DO 4 K=1,S
                H=F(K)
                DO 3 L=K,S
                     W(K,L)=W(K,L)+H*F(L)
    3           CONTINUE
    4      CONTINUE
    5 CONTINUE
      CALL LDLT (S,SDX,W,F,EPS1,IFLAG)
      IF(IFLAG.NE.0) GOTO 18
      DETW=1.
      DO 6 K=1,S
           DETW=DETW*W(K,K)
    6 CONTINUE
      IF(DETW.GE.EPS2) GOTO 7
      IFLAG=9
      GOTO 18
    7 IF(N.LE.1) GOTO 18
      IS=0
      I=0
```

*Table 9.9 – continued*

```
    8 IS=IS+1
      IF(IS.GT.M) GOTO 18
    9 I=I+1
      IF(I.LE.M) GOTO 10
      IT=IT+1
      IF(IT.GT.15) GOTO 18
      I=1
   10 P=Z(I)
      L=MJ(P)
      IF(L.LE.MO) GOTO 8
      ALPHAP= - L/(L-1.)
      DO 12 K=1,S
            H=X(I,K)
            XI(K)=H
            F(K)=XBAR(P,K)-H
            DO 11 L=1,K
                  WBAR(K,L)=W(K,L)
   11       CONTINUE
   12 CONTINUE
      CALL UPDATE (S,SDX,ALPHAP,F,WBAR,DET,EPS3,IFLAG)
      IF(IFLAG.NE.0) GOTO 18
      EQ=BIG
      DO 14 J=1,N
            CALL DISTW(XBAR,NDX,N,XI,WBAR,SDX,S,J,DIST,F)
            IF(J.NE.P) GOTO 13
            EP= -ALPHAP*DIST
            GOTO 14
   13       L=MJ(J)
            ALPHAJ= L/(L+1.)
            EJ=ALPHAJ*DIST
            IF(EJ.GE.EQ) GOTO 14
            EQ=EJ
            Q=J
            ALPHAQ=ALPHAJ
   14 CONTINUE
      IF(EQ.GE.EP*R) GOTO 8
      DO 15 K=1,S
            F(K)=XBAR(Q,K)-XI(K)
   15 CONTINUE
      CALL UPDATE (S,SDX,ALPHAQ,F,WBAR,DET,EPS3,IFLAG)
      IF(IFLAG.NE.0) GOTO 18
      IF(DET.GE.DETW) GOTO 8
      IS=0
      DETW=DET
      L=MJ(P)
      U=L
      K=MJ(Q)
      V=K
      FP=1./(U-1.)
      FQ=1./(V+1.)
      MJ(P)=L-1
      MJ(Q)=K+1
      DO 17 K=1,S
            T=XI(K)
            XBAR(P,K)=(U*XBAR(P,K)-T)*FP
            XBAR(Q,K)=(V*XBAR(Q,K)+T)*FQ
            DO 16 L=1,K
                  W(K,L)=WBAR(K,L)
   16       CONTINUE
   17 CONTINUE
      Z(I)=Q
      GOTO 9
   18 RETURN
      END
```

## 9.5 DWBEXM: Exchange method for the generalized criterion of adaptive distances

In the implementation of the exchange method for the criterion of adaptive distances, the scatter matrices $W_j$ for the starting partition are needed. For a fixed J, and using the mean vector (XBAR(J, K), K=1, S) determined with MEANS, the subroutine WJSCAT shown in Table 9.10 calculates the scatter matrix and stores it in ((WJ(K, L), L=K, S), K=1, S).

*Table 9.10*

```
      SUBROUTINE WJSCAT (X,MDX,M,S,SDX,Z,XBAR,NDX,N,J,WJ,F)
      INTEGER    Z(M),S,SDX
      DIMENSION X(MDX,S),XBAR(NDX,S),WJ(SDX,S),F(S)
      DO 1 K=1,S
      DO 1 L=K,S
    1 WJ(K,L)=0.
      DO 5 I=1,M
            L=Z(I)
            IF(L.NE.J) GOTO 5
            DO 2 K=1,S
                  F(K)=XBAR(J,K)-X(I,K)
    2       CONTINUE
            DO 4 K=1,S
                  H=F(K)
                  DO 3 L=K,S
                        WJ(K,L)=WJ(K,L)+H*F(L)
    3             CONTINUE
    4       CONTINUE
    5 CONTINUE
      RETURN
      END
```

In the subroutine DWBEXM shown in Table 9.11, the exchange method for the generalized objective function (4.15) is implemented with BETA $= \beta > 0$. If, at the start, $m_0 \leqslant s + 1$, the statement M0 = S + 1 is executed to satisfy the necessary condition for $C \in P^{++}(n,M)$. By choosing $m_0 > s + 1$, the user is given the opportunity to aim for 'sufficiently' positive definite scatter matrices. The computation of mean vectors, scatter matrices, and their determinants is only performed when $m_0\, n < m$; otherwise DWBEXM immediately returns with IFLAG = 3.

As in the previous section, the values det $W_j$ are calculated after corresponding calls to LDLT. Here EPS is set equal to ESP1, and if necessary, IFLAG = 4 is returned. If det $W_j <$ EPS2 for some $j=1, \ldots, n$, DWBEXM immediately returns with IFLAG = 5.

If the exchange condition (4.19) is satisfied for at least one $j \neq p$, the (first) subscript q is determined, for which the reduction in the objective function is greatest. In fact, the exchange is only performed when

$$(\det W_q')^\beta - (\det W_q)^\beta < R\,((\det W_p)^\beta - (\det W_p')^\beta) \qquad (9.19)$$

Again, R is set equal to .999. When updating, UPDATE, called with EPS = EPS3, can return IFLAG = 6 or IFLAG = 7, thus terminating DWBEXM prematurely.

After a regular exit (IFLAG = 0) of DWBEXM, (DETWJB (J), J=1, N) contains the values of (det $W_j)^\beta$, and D contains their sum for the end partition found. The arrays W, $W_j$, WP, WQ and F are used as workspace. Their size in the calling program must be adequate to accommodate the array bounds declared in the DIMENSION statements.

Usually, $\beta = 1/s$ is used. According to Chapter 4, all values $0 < \beta \leqslant 1$ are meaningful. Formally, all values $\beta > 0$ are permitted. However, with large values of $\beta$ ($\beta \approx 10$), one must expect cancellation errors in the subtraction of very large numbers. Furthermore, in practice one does not get plausible results, since, for large $\beta$, (5.47) is approached asymptotically.

*Table 9.11*

```
      SUBROUTINE DWBEXM (X,MDX,M,SDX,S,Z,MO,MJ,XBAR,NDX,N,
     *                   BETA,DETWJB,D,IT,IFLAG,W,WJ,WP,WQ,F)
      INTEGER   Z(M),MJ(N),S,SDX,P,Q,S1
      DIMENSION X(MDX,S),XBAR(NDX,S),DETWJB(N)
      DIMENSION W(NDX,SDX,S),WJ(SDX,S),WP(SDX,S),WQ(SDX,S),F(S)
      D=0.
      IT=0
      S1=S+1
      IF(MO.LE.S1) MO=S1
      IF(MO*N.LT.M) GOTO 1
      IFLAG=3
      GOTO 17
    1 BIG=1.E50
      R=.999
      EPS1=1.E-6
      EPS2=1.E-10
      EPS3=1.E-6
      CALL MEANS (X,MDX,M,S,Z,MO,MJ,XBAR,NDX,N,IFLAG)
      IF(IFLAG.NE.0) GOTO 17
      DO 5 J=1,N
            CALL WJSCAT (X,MDX,M,S,SDX,Z,XBAR,NDX,N,J,WJ,F)
            CALL LDLT (S,SDX,WJ,F,EPS1,IFLAG)
            IF(IFLAG.NE.0) GOTO 17
            H=1.
            DO 3 K=1,S
                  H=H*WJ(K,K)
                  DO 2 L=1,K
                        W(J,K,L)=WJ(K,L)
    2             CONTINUE
    3       CONTINUE
            IF(H.GE.EPS2) GOTO 4
            IFLAG=5
            GOTO 17
    4       IF(BETA.NE.1.) H=H**BETA
            DETWJB(J)=H
            D=D+H
    5 CONTINUE
      IF(N.LE.1) GOTO 17
      IS=0
      I=0
    6 IS=IS+1
      IF(IS.GT.M) GOTO 17
    7 I=I+1
      IF(I.LE.M) GOTO 8
      IT=IT+1
      IF(IT.GT.15) GOTO 17
      I=1
    8 P=Z(I)
      L=MJ(P)
```

*Table 9.11 (continued)*

```
      IF(L.LE.MO) GOTO 6
      V=L
      EQ=BIG
      DETBP=DETWJB(P)
      DO 14 J=1,N
            U=MJ(J)
            DO 10 K=1,S
                  F(K)=XBAR(J,K)-X(I,K)
                  DO 9 L=1,K
                        WJ(K,L)=W(J,K,L)
 9                CONTINUE
10          CONTINUE
            IF(J.EQ.P) ALPHA = - U/(U-1.)
            IF(J.NE.P) ALPHA =   U/(U+1.)
            CALL UPDATE (S,SDX,ALPHA,F,WJ,DETJ,EPS3,IFLAG)
            IF(IFLAG.NE.0) GOTO 17
            DETBJ=DETJ
            IF(BETA.NE.1.) DETBJ=DETJ**BETA
            IF(J.NE.P) GOTO 12
            EP=DETBP-DETBJ
            DETBP=DETBJ
            DO 11 K=1,S
            DO 11 L=1,K
11          WP(K,L)=WJ(K,L)
            GOTO 14
12          EJ=DETBJ-DETWJB(J)
            IF(EJ.GE.EQ) GOTO 14
            EQ=EJ
            DETBQ=DETBJ
            DO 13 K=1,S
            DO 13 L=1,K
13          WQ(K,L)=WJ(K,L)
            Q=J
            H=U
14    CONTINUE
      IF(EQ.GE.EP*R) GOTO 6
      IS=0
      DETWJB(P)=DETBP
      DETWJB(Q)=DETBQ
      D=D-EP+EQ
      DO 15 K=1,S
      DO 15 L=1,K
      W(P,K,L)=WP(K,L)
15    W(Q,K,L)=WQ(K,L)
      FP=1./(V-1.)
      FQ=1./(H+1.)
      DO 16 K=1,S
            T=X(I,K)
            XBAR(P,K)=(V*XBAR(P,K)-T)*FP
            XBAR(Q,K)=(H*XBAR(Q,K)+T)*FQ
16    CONTINUE
      Z(I)=Q
      MJ(P)=MJ(P)-1
      MJ(Q)=MJ(Q)+1
      GOTO 7
17    RETURN
      END
```

**Problem 9.1**: How would you implement the minimal distance method for the determinant criterion and the criterion of adaptive distances? Which of the subroutines MEANS, LDLT, UPDATE, DISTW and WJSCAT could you use as building blocks?

**Problem 9.2**: How would you need to change TRWEXM, DETEXM and DWBEXM in order to implement a method that performed an exchange as soon as the exchange condition was satisfied, as discussed in Section 2.8?

**Problem 9.3**: Change DETEXM in such a way that the exchange condition is implemented in the form of (3.44) or (3.51). Examine closely the assertion made in the text that the number of computations required per exchange attempt is of the same order of magnitude as that using condition (3.49).

**Problem 9.4** (continuation of Problem 5.2): Implement the exchange formula developed there for the various objective functions.

# 10 Implementations for the $L_1$ criterion with different data types, for criteria not involving centres, and for clusterwise regression analysis

## 10.1 OVSEXM, OVREXM: Exchange methods for the $L_1$ criterion

In this section we describe two subroutines OVSEXM and OVREXM, shown in Tables 10.1 and 10.2, which perform the exchange method for the $L_1$ criterion (6.2). They contain the sorting necessary to determine the median vectors for the initial partition (DO loops numbers 5 and 6). The main difference is that OVSEXM stores the components of the data matrix ((X(I, K), K=1, S), I=1, M), which have been sorted cluster by cluster to find the medians, in a work space array ((XX(I, K), K=1, S), I=1, M) of the same size as X. The values of XX are set in DO loop number 8. They are later directly available and can be updated. This is not the case with OVREXM, which therefore needs more computation time. Although the structure of both routines is identical to that of the routines described previously, the two programs are rather longer because of the need to distinguish different cases during up- and downdating processes (6.34) to (6.42).

Both programs are written for data matrices with arbitrary integer, not real, values, though they can easily be converted, as will be described later. One reason for this is to allow a comparison with the computer time needed for a third subroutine OVPEXM. The latter is designed especially for ordinal data matrices, corresponding to Section 6.7, and will be described in the following section. Moreover, in practical applications such as the analysis of questionnaires or the classification of time series (for example, demand per week, number of orders per month), the data matrices do often contain integers. For these, integer arithmetic is in general faster than floating point arithmetic, at least for addition and subtraction. (On the TR 440, the ratio of integer to floating point addition times is 2:7.)

To match the integer data matrix, the subroutine parameters U, V, E and D are also of integer type and must be declared as such. ((U(J, K), V(J, K), K=1, S), J=1,N) are the median vectors. Note that, for odd MJ(J), U(J,K) equals V(J, K). (E(J), J=1, N) are the measures of compactness, and their sum D is the value of the objective function for the end partition. Similarly, the workspace arrays Y, UP, VP, UJ, VJ, UQ, VQ required by both subroutines are integer. The LOGICAL array EVEN is also used in both programs. In addition, OVSEXM uses the integer workspace arrays XX, SP, SJ, SQ and MK. MK has dimension

N1 which must be set equal to N + 1 in the calling program. Further parameters for the two routines have the same meaning as before.

Where the FORTRAN compiler and the size of the input and output data (for example questionnaire data) permit it, all integer arrays could be declared as INTEGER * 2, thereby almost halving the storage required.

*Table 10.1*

```
      SUBROUTINE OVSEXM (X,MDX,M,S,Z,MJ,NDX,N,U,V,E,D,
     *                   IT,IFLAG,Y,EVEN,UP,VP,UJ,VJ,UQ,VQ,
     *                   SP,SJ,SQ,N1,MK,XX)
      INTEGER S,Z(M),MJ(N),E(N),D,
     *        X(MDX,S),U(NDX,S),V(NDX,S),XX(MDX,S)
      INTEGER Y(M),UP(S),VP(S),UJ(S),VJ(S),UQ(S),VQ(S),
     *        G,SP(S),SJ(S),SQ(S),MK(N1),
     *        P,Q,R,R1,H,F,EP,EJ,EQ,YR,YR1,XU,XV,
     *        UPK,VPK,UJK,VJK,XIK,XLK
      LOGICAL EVEN(N),EVENP,FINAL,REVERS
      KBIG=99999999
      IT=0
      D=0
      MO=1
      DO 1 J=1,N
            MJ(J)=0
            E(J)=0
    1 CONTINUE
      IFLAG=1
      DO 2 I=1,M
            J=Z(I)
            IF(J.LT.1.OR.J.GT.N) GOTO 38
            MJ(J)=MJ(J)+1
    2 CONTINUE
      IFLAG=0
      MK(1)=1
      DO 10 J=1,N
            L=MJ(J)
            IF(L.GE.MO) GOTO 3
            IFLAG=2
            GOTO 38
    3       L1=L/2
            L2=(L+1)/2
            EVEN(J)=(2*L1).EQ.L
            LJ=MK(J)
            MK(J+1)=LJ+L
            DO 9 K=1,S
                  R=0
                  DO 4 I=1,M
                        IF(Z(I).NE.J) GOTO 4
                        R=R+1
                        Y(R)=X(I,K)
    4             CONTINUE
                  IF(L.EQ.1) GOTO 7
                  DO 6 P=2,L
                        FINAL=.TRUE.
                        Q=L-P+1
                        DO 5 R=1,Q
                              R1=R+1
                              YR=Y(R)
                              YR1=Y(R1)
                              IF(YR.LE.YR1) GOTO 5
                              Y(R1)=YR
                              Y(R)=YR1
                              FINAL=.FALSE.
    5                   CONTINUE
                        IF(FINAL) GOTO 7
    6             CONTINUE
```

*Table 10.1 – continued*

```
  7              H=Y(L2)
                 U(J,K)=H
                 IF(EVEN(J)) H=Y(L2+1)
                 V(J,K)=H
                 F=0
                 DO 8 R=1,L
                       YR=Y(R)
                       MR=LJ+R-1
                       XX(MR,K)=YR
                       F=F+IABS(YR-H)
  8              CONTINUE
                 E(J)=E(J)+F
                 D=D+F
  9        CONTINUE
 10   CONTINUE
      IF(N.LE.1) GOTO 38
      IS=0
      I=0
 11   IS=IS+1
      IF(IS.GT.M) GOTO 38
 12   I=I+1
      IF(I.LE.M) GOTO 13
      IT=IT+1
      IF(IT.GT.15) GOTO 38
      I=1
 13   P=Z(I)
      R=MJ(P)
      IF(R.LE.MO) GOTO 11
      MP=MK(P)
      G=(R+1)/2+MP-1
      EP=0
      EVENP=EVEN(P)
      DO 20 K=1,S
            UPK=U(P,K)
            VPK=V(P,K)
            XIK=X(I,K)
            DO 14 MR=1,R
                        MS=R-MR+MP
                        IF(XIK.NE.XX(MS,K)) GOTO 14
                        SP(K)=MS
                        GOTO 15
 14         CONTINUE
 15         IF(EVENP) GOTO 18
            XU=XX(G-1,K)
            XV=XX(G+1,K)
            IF(XIK.NE.UPK) GOTO 16
            UP(K)=XU
            VP(K)=XV
            GOTO 20
 16         IF(XIK.GT.UPK) GOTO 17
            UP(K)=UPK
            VP(K)=XV
            EP=EP+IABS(XIK-UPK)
            GOTO 20
 17         UP(K)=XU
            VP(K)=UPK
            EP=EP+IABS(XIK-UPK)
            GOTO 20
 18         IF(XIK.GT.UPK) GOTO 19
            UP(K)=VPK
            VP(K)=VPK
            EP=EP+IABS(XIK-VPK)
            GOTO 20
 19         UP(K)=UPK
            VP(K)=UPK
            EP=EP+IABS(XIK-UPK)
```

*Table 10.1 – continued*

```
20 CONTINUE
   EQ=KBIG
   DO 29 J=1,N
         IF(J.EQ.P) GOTO 29
         R=MJ(J)
         EJ=0
         MR=MK(J)
         FINAL=R.GT.1
         G=(R+1)/2+MR-1
         MS=MK(J+1)
         DO 27 K=1,S
               UJK=U(J,K)
               VJK=V(J,K)
               XIK=X(I,K)
               MT=MS
               DO 21 L=1,R
                     MP=MR+L-1
                     IF(XIK.GT.XX(MP,K)) GOTO 21
                     MT=MP
                     GOTO 22
21             CONTINUE
22             SJ(K)=MT
               IF(EVEN(J)) GOTO 24
               IF(XIK.GT.UJK) GOTO 23
               UJ(K)=XIK
               XU=XIK
               IF(FINAL) XU=XX(G-1,K)
               IF(XIK.LT.XU) UJ(K)=XU
               VJ(K)=VJK
               EJ=EJ+IABS(XIK-UJK)
               GOTO 27
23             VJ(K)=XIK
               UJ(K)=UJK
               XV=XIK
               IF(FINAL) XV=XX(G+1,K)
               IF(XIK.GT.XV) VJ(K)=XV
               EJ=EJ+IABS(XIK-UJK)
               GOTO 27
24             IF(XIK.GE.VJK) GOTO 26
               IF(XIK.LE.UJK) GOTO 25
               UJ(K)=XIK
               VJ(K)=XIK
               GOTO 27
25             UJ(K)=UJK
               VJ(K)=UJK
               EJ=EJ+IABS(XIK-UJK)
               GOTO 27
26             UJ(K)=VJK
               VJ(K)=VJK
               EJ=EJ+IABS(XIK-VJK)
27       CONTINUE
         IF(EJ.GE.EQ) GOTO 29
         EQ=EJ
         Q=J
         DO 28 K=1,S
               UQ(K)=UJ(K)
               VQ(K)=VJ(K)
               SQ(K)=SJ(K)
28       CONTINUE
29 CONTINUE
   IF(EQ.GE.EP) GOTO 11
   E(P)=E(P)-EP
   E(Q)=E(Q)+EQ
   D=D-EP+EQ
   MJ(P)=MJ(P)-1
```

*Table 10.1 – continued*

```
      MJ(Q)=MJ(Q)+1
      EVEN(P)=.NOT.EVEN(P)
      EVEN(Q)=.NOT.EVEN(Q)
      REVERS=P.GT.Q
      DO 33 K=1,S
            U(P,K)=UP(K)
            V(P,K)=VP(K)
            U(Q,K)=UQ(K)
            V(Q,K)=VQ(K)
            XIK=X(I,K)
            MP=SP(K)
            MQ=SQ(K)
            IF(REVERS) GOTO 31
            MQ1=MQ-1
            IF(MP.EQ.MQ1) GOTO 33
            MQ2=MQ1-1
            DO 30 L=MP,MQ2
                  XX(L,K)=XX(L+1,K)
30          CONTINUE
            XX(MQ1,K)=XIK
            GOTO 33
31          MQ2=MQ+1
            H=XX(MQ,K)
            XX(MQ,K)=XIK
            DO 32 L=MQ2,MP
                  F=XX(L,K)
                  XX(L,K)=H
                  H=F
32          CONTINUE
33    CONTINUE
      IF(REVERS) GOTO 35
      G=P+1
      DO 34 L=G,Q
            MK(L)=MK(L)-1
34    CONTINUE
      GOTO 37
35    G=Q+1
      DO 36 L=G,P
            MK(L)=MK(L)+1
36    CONTINUE
37    Z(I)=Q
      IS=0
      GOTO 12
38    RETURN
      END
```

As indicated, the changes needed to handle real data matrices are quite simple. In OVSEXM, the two INTEGER statements would need to be replaced by

```
      INTEGER S, Z(M), MJ(N), SP(S), SJ(S), SQ(S),
  *         MK (N1), G, P, Q, R, R1
      DIMENSION X(MDX, S), U(NDX, S), V(NDX, S), E(N),
  *         XX(MDX, S), Y(M), UP(S), VP(S),
  *         UJ(S), VJ(S), UQ(S), VQ(S)
```

The statement

KBIG = 99999999

(KBIG plays the same role as BIG did previously) would need to be replaced by

REAL KBIG

KBIG = 1.E50

The IABS function would need to be replaced by the ABS function. Overall, and ignoring differences in notation, one would end up with the original version published in [B39]. Because of rounding errors (which cannot occur under integer arithmetic) it would, in addition, be necessary to introduce a variable RR with the statement

RR = .999

and to replace the statement following statement number 29 by

IF(EQ.GE.EP*RR) GOTO 11 (10.1)

The need for such a change is discussed in detail in the description of TRWEXM.

*Table 10.2*

```
      SUBROUTINE OVREXM (X,MDX,M,S,Z,MJ,NDX,N,U,V,E,D,
     *                   IT,IFLAG,Y,EVEN,UP,VP,UJ,VJ,UQ,VQ)
      INTEGER S,Z(M),MJ(N),E(N),D,
     *        X(MDX,S),U(NDX,S),V(NDX,S)
      INTEGER Y(M),UP(S),VP(S),UJ(S),VJ(S),UQ(S),VQ(S),
     *        P,Q,R,R1,H,F,EP,EJ,EQ,YR,YR1,XU,XV,
     *        UPK,VPK,UJK,VJK,XIK,XLK
      LOGICAL EVEN(N),EVENP,FINAL
      KBIG=99999999
      IT=0
      D=0
      MO=1
      DO 1 J=1,N
           MJ(J)=0
           E(J)=0
    1 CONTINUE
      IFLAG=1
      DO 2 I=1,M
           J=Z(I)
           IF(J.LT.1.OR.J.GT.N) GOTO 33
           MJ(J)=MJ(J)+1
    2 CONTINUE
      IFLAG=0
      DO 10 J=1,N
           L=MJ(J)
           IF(L.GE.MO) GOTO 3
           IFLAG=2
           GOTO 33
    3      L1=L/2
           L2=(L+1)/2
           EVEN(J)=(2*L1).EQ.L
           DO 9 K=1,S
                R=0
```

*Table 10.2 – continued*

```
                DO 4 I=1,M
                     IF(Z(I).NE.J) GOTO 4
                     R=R+1
                     Y(R)=X(I,K)
  4             CONTINUE
                IF(L.EQ.1) GOTO 7
                DO 6 P=2,L
                     FINAL=.TRUE.
                     Q=L-P+1
                     DO 5 R=1,Q
                          R1=R+1
                          YR=Y(R)
                          YR1=Y(R1)
                          IF(YR.LE.YR1) GOTO 5
                          Y(R1)=YR
                          Y(R)=YR1
                          FINAL=.FALSE.
  5                  CONTINUE
                     IF(FINAL) GOTO 7
  6             CONTINUE
  7             H=Y(L2)
                U(J,K)=H
                IF(EVEN(J)) H=Y(L2+1)
                V(J,K)=H
                F=0
                DO 8 R=1,L
                     F=F+IABS(Y(R)-H)
  8             CONTINUE
                E(J)=E(J)+F
                D=D+F
  9        CONTINUE
 10   CONTINUE
      IF(N.LE.1) GOTO 33
      IS=0
      I=0
 11   IS=IS+1
      IF(IS.GT.M) GOTO 33
 12   I=I+1
      IF(I.LE.M) GOTO 13
      IT=IT+1
      IF(IT.GT.15) GOTO 33
      I=1
 13   P=Z(I)
      R=MJ(P)
      IF(R.LE.MO) GOTO 11
      R1=R/2
      EP=0
      EVENP=EVEN(P)
      DO 20 K=1,S
           UPK=U(P,K)
           VPK=V(P,K)
           XIK=X(I,K)
           IF(EVENP) GOTO 18
           LU=0
           LV=0
           XU= - KBIG
           XV=KBIG
           DO 15 L=1,M
                IF(Z(L).NE.P.OR.L.EQ.I) GOTO 15
                XLK=X(L,K)
                IF(XLK.EQ.UPK) GOTO 15
                IF(XLK.GT.UPK) GOTO 14
                XU=MAXO(XLK,XU)
                LU=LU+1
                GOTO 15
```

*Table 10.2 – continued*

```
14          XV=MINO(XLK,XV)
            LV=LV+1
15        CONTINUE
          IF(LU.LT.R1) XU=UPK
          IF(LV.LT.R1) XV=UPK
          IF(XIK.NE.UPK) GOTO 16
          UP(K)=XU
          VP(K)=XV
          GOTO 20
16        IF(XIK.GT.UPK) GOTO 17
          UP(K)=UPK
          VP(K)=XV
          EP=EP+IABS(XIK-UPK)
          GOTO 20
17        UP(K)=XU
          VP(K)=UPK
          EP=EP+IABS(XIK-UPK)
          GOTO 20
18        IF(XIK.GT.UPK) GOTO 19
          UP(K)=VPK
          VP(K)=VPK
          EP=EP+IABS(XIK-VPK)
          GOTO 20
19        UP(K)=UPK
          VP(K)=UPK
          EP=EP+IABS(XIK-UPK)
20    CONTINUE
      EQ=KBIG
      DO 31 J=1,N
          IF(J.EQ.P) GOTO 31
          R=MJ(J)
          R1=R/2
          EJ=0
          DO 29 K=1,S
              UJK=U(J,K)
              VJK=V(J,K)
              XIK=X(I,K)
              IF(EVEN(J)) GOTO 26
              LR=0
              IF(XIK.GT.UJK) GOTO 23
              UJ(K)=XIK
              IF(XIK.EQ.UJK.OR.R.EQ.1) GOTO 22
              XU= - KBIG
              DO 21 L=1,M
                  IF(Z(L).NE.J) GOTO 21
                  XLK=X(L,K)
                  IF(XLK.GE.UJK) GOTO 21
                  LR=LR+1
                  XU=MAXO(XLK,XU)
21            CONTINUE
              IF(LR.LT.R1) XU=UJK
              IF(XIK.LT.XU) UJ(K)=XU
22            VJ(K)=VJK
              EJ=EJ+IABS(XIK-UJK)
              GOTO 29
23            VJ(K)=XIK
              IF(R.EQ.1) GOTO 25
              XV=KBIG
              DO 24 L=1,M
                  IF(Z(L).NE.J) GOTO 24
                  XLK=X(L,K)
                  IF(XLK.LE.VJK) GOTO 24
                  LR=LR+1
                  XV=MINO(XLK,XV)
```

*Table 10.2 – continued*

```
24          CONTINUE
            IF(LR.LT.R1) XV=VJK
            IF(XIK.GT.XV) VJ(K)=XV
25          UJ(K)=UJK
            EJ=EJ+IABS(XIK-UJK)
            GOTO 29
26          IF(XIK.GE.VJK) GOTO 28
            IF(XIK.LE.UJK) GOTO 27
            UJ(K)=XIK
            VJ(K)=XIK
            GOTO 29
27          UJ(K)=UJK
            VJ(K)=UJK
            EJ=EJ+IABS(XIK-UJK)
            GOTO 29
28          UJ(K)=VJK
            VJ(K)=VJK
            EJ=EJ+IABS(XIK-VJK)
29       CONTINUE
         IF(EJ.GE.EQ) GOTO 31
         EQ=EJ
         Q=J
         DO 30 K=1,S
            UQ(K)=UJ(K)
            VQ(K)=VJ(K)
30       CONTINUE
31 CONTINUE
   IF(EQ.GE.EP) GOTO 11
   E(P)=E(P)-EP
   E(Q)=E(Q)+EQ
   D=D-EP+EQ
   MJ(P)=MJ(P)-1
   MJ(Q)=MJ(Q)+1
   EVEN(P)=.NOT.EVEN(P)
   EVEN(Q)=.NOT.EVEN(Q)
   DO 32 K=1,S
         U(P,K)=UP(K)
         V(P,K)=VP(K)
         U(Q,K)=UQ(K)
         V(Q,K)=VQ(K)
32 CONTINUE
   Z(I)=Q
   IS=0
   GOTO 12
33 RETURN
   END
```

In OVREXM, the two INTEGER statements would similarly have to be replaced, this time by

```
      INTEGER S, Z(M), MJ(N), P, Q, R, R1
      DIMENSION X(MDX, S), U(NDX, S), V(NDX, S), E(N),
     *          Y(M), UP(S), VP(S), UJ(S), VJ(S), UQ(S), VQ(S)
```

The other changes necessary are as described for OVSEXM above. The test (10.1) would have to replace the statement following statement number 31. In addition, the intrinsic functions MIN0 and MAX0 would have to be replaced by AMIN1 and AMAX1.

With both integer (present version) and real data matrices, both subroutines normally return with IFLAG = 0. The possible error indicates IFLAG = 1 and IFLAG = 2 refer to the initial partition and have the same meaning as described earlier for MEANS.

### 10.2 OVPEXM: Exchange method for the $L_1$ criterion with ordinal data

Subroutine OVPEXM, as listed in Table 10.3, is designed for a slightly more specialized situation than that suggested in (6.60). In the first instance, it is required that

$$x_i \in \{1, \ldots, t\}^s, \quad t \geqslant 2, \quad i=1, \ldots, m \qquad (10.2)$$

This situation often arises when handling responses to multiple choice questions where the replies are coded on a bipolar scale from 1 to t. Often, t = 7. Thus, OVPEXM expects an integer data matrix with

$$1 \leqslant x_{ik} \leqslant t, \quad i=1, \ldots, m, k=1, \ldots, s. \qquad (10.3)$$

Condition (10.3) is not tested, but the subroutine returns instantly with the error IFLAG = 13, when T < 2.

As always, OVPEXM tests the feasibility of the initial partition and, if necessary, breaks off with IFLAG = 1 or IFLAG = 2. Then the routine determines how many of the $k^{th}$ components of the vectors $x_i$ with $i \in C_j$ are equal to *l*, and the counts are stored in (((A(L, K, J), J=1, N), K=1, S), L=1, T). Compared with OVSEXM and OVREXM, far less workspace is needed: only the arrays B and C. For small values of t, OVPEXM may even use less storage. As before, all arrays may be declared as INTEGER * 2 if the compiler and the data values allow it.

*Table 10.3*

```
      SUBROUTINE OVPEXM (X,MDX,M,SDX,S,Z,MJ,NDX,N,E,D,IT,
     *                   A,TDX,T,IFLAG,B,C)
      INTEGER S,SDX,TDX,T,Z(M),MJ(N),E(N),D
      INTEGER X(MDX,S),A(TDX,SDX,N),B(T),C(N),
     *        F,P,Q,EJ,EQ,EP,BIG,XIK,SUM
      IT=0
      D=0
      IFLAG=13
      IF(T.LT.2) GOTO 20
      IFLAG=0
      MO=1
      BIG=99999999
      DO 2 J=1,N
           MJ(J)=0
           E(J)=0
           DO 1 K=1,S
           DO 1 L=1,T
    1      A(L,K,J)=0
```

*Table 10.3 – continued*

```
    2 CONTINUE
      DO 6 I=1,M
            J=Z(I)
            IF(J.GE.1.AND.J.LE.N) GOTO 3
            IFLAG=1
            GOTO 20
    3       MJ(J)=MJ(J)+1
            DO 5 K=1,S
                  XIK=X(I,K)
                  DO 4 L=1,T
                        IF(XIK.EQ.L) A(L,K,J)=A(L,K,J)+1
    4             CONTINUE
    5       CONTINUE
    6 CONTINUE
      DO 10 J=1,N
            IF(MJ(J).GE.MO) GOTO 7
            IFLAG=2
            GOTO 20
    7       F=0
            DO 9 K=1,S
                  DO 8 L=1,T
                        B(L)=A(L,K,J)
    8             CONTINUE
                  CALL MEDIAN (B,T,SUM)
                  F=F+SUM
    9       CONTINUE
            D=D+F
            E(J)=F
   10 CONTINUE
      IF(N.LE.1) GOTO 20
      IS=0
      I=0
   11 IS=IS+1
      IF(IS.GT.M) GOTO 20
   12 I=I+1
      IF(I.LE.M) GOTO 13
      IT=IT+1
      IF(IT.GT.15) GOTO 20
      I=1
   13 P=Z(I)
      IF(MJ(P).LE.MO) GOTO 11
      DO 16 J=1,N
            C(J)=0
            DO 15 K=1,S
                  XIK=X(I,K)
                  DO 14 L=1,T
                        B(L)=A(L,K,J)
                        IF(XIK.NE.L) GOTO 14
                        IF(J.EQ.P) B(L)=B(L)-1
                        IF(J.NE.P) B(L)=B(L)+1
   14             CONTINUE
                  CALL MEDIAN (B,T,SUM)
                  C(J)=C(J)+SUM
   15       CONTINUE
   16 CONTINUE
      EQ= - BIG
      EP=C(P)-E(P)
      DO 17 J=1,N
            IF(J.EQ.P) GOTO 17
            EJ=E(J)-C(J)
            IF(EJ.LE.EQ) GOTO 17
            EQ=EJ
            Q=J
```

*Table 10.3 –continued*

```
   17 CONTINUE
      IF(EP.GE.EQ) GOTO 11
      IS=0
      D=D+EP-EQ
      E(P)=C(P)
      E(Q)=C(Q)
      MJ(P)=MJ(P)-1
      MJ(Q)=MJ(Q)+1
      DO 19 K=1,S
            XIK=X(I,K)
            DO 18 L=1,T
                  IF(XIK.NE.L) GOTO 18
                  A(L,K,P)=A(L,K,P)-1
                  A(L,K,Q)=A(L,K,Q)+1
   18       CONTINUE
   19 CONTINUE
      Z(I)=Q
      GOTO 12
   20 RETURN
      END
```

The alternative form of the objective function (6.71) requires the quantities $f_{r(j,k)}$. These are determined in subroutine MEDIAN shown in Table 10.4. In the first DO loop, $f_1$ is calculated as in (6.66), and in the second, the $f_l$ ($l=2, \ldots, r-1$) are calculated up to the first subscript r for which $g_r = U-V \geqslant 0$ holds. This is done by a call to MEDIAN, first for the numbers (((A(L, K, J), J = 1, N), K = 1, S), L = 1, T) calculated for the initial partition, and then for the corresponding numbers after they have been modified to take account of the removal or addition of a vector $x_i$. After a successful exchange, the A(L, K, J) are updated.

*Tabelle 10.4*

```
      SUBROUTINE MEDIAN (B,T,SUM)
      INTEGER T,B(T),BL,H,Z,U,V,SUM
      H=0
      Z=0
      DO 1 L=2,T
            I=T-L+2
            H=H+B(I)
            Z=Z+H
    1 CONTINUE
      V=B(1)
      U=H
      DO 2 L=2,T
            IF(U.LE.V) GOTO 3
            Z=Z+V-U
            BL=B(L)
            U=U-BL
            V=V+BL
    2 CONTINUE
    3 SUM=Z
      RETURN
      END
```

Provided that IFLAG = 0, OVPEXM returns the same end partition as OVSEXM and OVREXM, given the same starting partition. In Section 12.1 we shall see which of the three subroutines is to be preferred, depending on the values of m, n, s and t.

If one wants to change OVPEXM to allow the use of ordinal vectors as in (6.60), that is, if one wants to permit a data matrix with

$$1 \leqslant x_{ik} \leqslant t_k, \ t_k \geqslant 2, \quad i = 1, \ldots, m, \ k = 1, \ldots, s \tag{10.4}$$

instead of (10.3), the following simple changes need to be made.

In the calling program, the first dimension of the array A, currently TDX, that is, the maximum value allowed for t, would have to be $\max_k t_k$ and this value would also have to be passed to OVPEXM.

The loops of the form DO ... L=1, T would have to be preceded by an extra statement KT = T(K), the DO statements would have to be changed to DO ... L=1, KT, and inside the loops, the variable T would have to be replaced by KT. In addition, the array T would have to be declared with dimension S, and the correct values (T(K), K=1, S) would have to be passed as parameters. In MEDIAN as well as in OVPEXM, the work space array B would need to have dimension TDX rather than T. Finally, the statement

```
IF (T.LT.2) GOTO 20
```

would have to be replaced by the three statements

```
   DO 21 K=1, S
         IF (T(K).LT.2) GOTO 20
21 CONTINUE
```

## 10.3 BVPEXM: Exchange method for the $L_1$ criterion with binary data

Subroutine BVPEXM, listed in Table 10.5, implements the $L_1$ criterion (6.55) for a binary data matrix, as described in Section 6.6. Since the elements of the data matrix consist of zeros and ones only, these may be coded as .FALSE. and .TRUE. or F and T. The subroutine is written on the assumption that the data matrix is coded in this way in a LOGICAL array. This has the advantage that many FORTRAN compilers allow the statement

```
LOGICAL X(MDX, S), XIK
```

at the beginning of BVPEXM, and the corresponding declaration of X in the calling program, to be replaced by

LOGICAL*1 X(MDX, S), XIK (10.5)

This reduces the storage required for X by 75%, a worthwhile saving.

The other parameters of BVPEXM leave the same meaning as for OVPEXM, and are of the same data type.

The array ((A(J, K), K=1, S), J=1, N) now only has two dimensions. It is used for the number of ones or .TRUE.s in the $k^{th}$ compound of the binary vectors belonging to the $j^{th}$ cluster. Incidentally, it would be easy to use (6.54) to calculate from these numbers for the end partition the corresponding class median vectors.

*Table 10.5*

```
      SUBROUTINE BVPEXM (X,MDX,M,S,Z,MJ,NDX,N,E,D,IT,IFLAG,A,C)
      INTEGER S,D,Z(M),MJ(N),E(N)
      INTEGER A(NDX,S),C(N),AJK,F,P,Q,Y,EJ,EQ,EP,BIG
      LOGICAL X(MDX,S),XIK
      IT=0
      D=0
      MO=1
      BIG=99999999
      IFLAG=0
      DO 2 J=1,N
           MJ(J)=0
           E(J)=0
           DO 1 K=1,S
                A(J,K)=0
 1         CONTINUE
 2    CONTINUE
      DO 5 I=1,M
           J=Z(I)
           IF(J.GE.1.AND.J.LE.N) GOTO 3
           IFLAG=1
           GOTO 19
 3         MJ(J)=MJ(J)+1
           DO 4 K=1,S
                IF(X(I,K)) A(J,K)=A(J,K)+1
 4         CONTINUE
 5    CONTINUE
      DO 8 J=1,N
           L=MJ(J)
           IF(L.GE.MO) GOTO 6
           IFLAG=2
           GOTO 19
 6         F=0
           DO 7 K=1,S
                AJK=A(J,K)
                F=F+MINO(AJK,L-AJK)
 7         CONTINUE
           D=D+F
           E(J)=F
 8    CONTINUE
      IF(N.LE.1) GOTO 19
      IS=0
      I=0
 9    IS=IS+1
      IF(IS.GT.M) GOTO 19
10    I=I+1
      IF(I.LE.M) GOTO 11
      IT=IT+1
      IF(IT.GT.15) GOTO 19
      I=1
```

*Table 10.5 – continued*

```
11 P=Z(I)
   L=MJ(P)
   IF(L.LE.MO) GOTO 9
   DO 12 J=1,N
        C(J)=0
12 CONTINUE
   DO 16 K=1,S
        XIK=X(I,K)
        DO 15 J=1,N
             Y=A(J,K)
             L=MJ(J)
             IF(J.EQ.P) GOTO 13
             L=L+1
             IF(XIK) Y=Y+1
             GOTO 14
13           L=L-1
             IF(XIK) Y=Y-1
14           C(J)=C(J)+MINO(Y,L-Y)
15      CONTINUE
16 CONTINUE
   EQ= - BIG
   EP=C(P)-E(P)
   DO 17 J=1,N
        IF(J.EQ.P) GOTO 17
        EJ=E(J)-C(J)
        IF(EJ.LE.EQ) GOTO 17
        EQ=EJ
        Q=J
17 CONTINUE
   IF(EP.GE.EQ) GOTO 9
   IS=0
   D=D+EP-EQ
   E(P)=C(P)
   E(Q)=C(Q)
   MJ(P)=MJ(P)-1
   MJ(Q)=MJ(Q)+1
   DO 18 K=1,S
        IF(.NOT.X(I,K)) GOTO 18
        A(P,K)=A(P,K)-1
        A(Q,K)=A(Q,K)+1
18 CONTINUE
   Z(I)=Q
   GOTO 10
19 RETURN
   END
```

Where the data matrix can be declared as LOGICAL*1 with a 75% saving on storage, it is worth considering whether ordinal data matrices should be converted into binary form in some suitable way, so that the storage requirement can be reduced by using BVPEXM rather than OVPEXM. Indeed, a data matrix such as (10.4) can be mapped one-to-one onto a binary matrix $\tilde{X}$, using

$$x_{ik} = \ell \iff \tilde{x}_{i,p_k+r} = \begin{cases} 0 \text{ for } r = 1, \ldots, t_k - \ell \\ 1 \text{ for } r = t_k - \ell + 1, \ldots, t_k - 1, \end{cases} \tag{10.6}$$

$$\text{where } p_1 := 0,\ p_{k+1} := p_k + t_k - 1 \ (k=1, \ldots, s-1),$$

The number of columns in $\tilde{X}$ is

$$\tilde{s} = \sum_{k=1}^{s} (t_k - 1) = \sum_{k=1}^{s} t_k - s \tag{10.7}$$

For example, let $s = 3$, $t_1 = 3$, $t_2 = 5$, $t_3 = 2$. The above transformation converts the ordinal vector $x = (2, 3, 1)^T$ into the binary vector $\tilde{x} = (0, 1, 0, 0, 1, 1, 0)^T$ and vice versa. The property which makes the transformation interesting in the present context is the invariance of the norm:

$$\| x - y \|_1 = \| \tilde{x} - \tilde{y} \|_1 \, . \tag{10.8}$$

Because

$$\| x - y \|_1 = \sum_{k=1}^{s} | x_k - y_k | ,$$

(10.8) can be proved by demonstrating the equality for the $k^{th}$ component. Let $x_k = \ell_1$ and $y_k = \ell_2$. For the corresponding portions $\tilde{x}_k$ and $\tilde{y}_k$ of $\tilde{x}$ and $\tilde{y}$ we have

$$\tilde{x}_k = (0, 0, 0, 0, \ldots, 1, 1, 1)^T \quad \text{with } t_k - \ell_1 \text{ zeros}$$

$$\tilde{y}_k = (0, 0, \ldots, 1, 1, 1, 1, 1)^T \quad \text{with } t_k - \ell_2 \text{ zeros}$$

Therefore,

$$| x_k - y_k | = | \ell_1 - \ell_2 | = \| \tilde{x}_k - \tilde{y}_k \|_1 \, .$$

If we denote the class median vectors for X by $y_j$, and those for $\tilde{X}$ by $\tilde{y}_j$, then (10.8) implies

$$\sum_{j=1}^{n} \min_{y_j \in \mathbb{R}^s} \sum_{i \in C_j} \| x_i - y_j \|_1 = \sum_{j=1}^{n} \min_{\tilde{y}_j \in \mathbb{R}^{\tilde{s}}} \sum_{i \in C_j} \| \tilde{x}_i - \tilde{y}_j \|_1 \, . \tag{10.9}$$

for all partitions $C \in P(n, M)$. Thus, the $L_1$ criterion is identical for both forms of the data matrix, and the use of OVPEXM with X is equivalent to the use of BVPEXM with $\tilde{X}$.

Under the above conditions, there is a saving on storage if $\tilde{s} < 4s$. Thus, for $t_k = 3$, the saving is 50%. For $t_k = 4$, it is 25%. For $t_k = 5$, the storage requirements are equal. For $\tilde{s} \geqslant 4s$ the amount of computation needed with BVPEXM could well be less than with OVPEXM since it avoids the use of subroutine MEDIAN which takes a lot of time.

The transformation (10.6) can also be useful in a totally different context If a given data matrix contains a mixture of binary and ordinal variables then one can transform every ordinal characteristic with $t_k$ possible values into $t_k - 1$ binary characteristics while retaining the information content, and then apply BVPEXM. Alternatively, it may be sensible to regard the binary characteristics as ordinal characteristics with values 1 and 2, and to process the ordinal matrix with OVPEXM or even OVSEXM or OVREXM. The user must decide whether this should be done in a given case.

## 10.4 TIHEXM: Exchange method for three centre-free criteria

We now consider the approximate minimization of the objective functions $M_1$, $M_2$ and $M_3$ in Section 7.1, using the exchange method. To save storage, the distance matrix $(t_{ih})$ is not declared as a square matrix. Instead, the lower half of $(t_{ih})$ is stored contiguously without the diagonal in a one-dimensional array (T(K), K=1, M12) of size M12 = (m(m−1))/2. For m = 4, the assignment may be made as follows:

$$\begin{matrix} t_{21} & & & & T(1) & & \\ t_{31} & t_{32} & & \rightarrow & T(2) & T(3) & \\ t_{41} & t_{42} & t_{43} & & T(4) & T(5) & T(6)\,. \end{matrix}$$

In general, the assignment of storage is given by

$$T(K) = t_{ih},\ \ K = ((i-1)(i-2))/2 + h,\ \ (i=2, \ldots, m,\ h=1, \ldots, i-1) \quad (10.10)$$

Where the exchange formulas in Section 7.1 require a matrix element $t_{ih} = t_{hi}$ $(i \neq h)$, the subscript K for the array (T(K), K=1, M12) is found via

$$\begin{aligned} i > h \ &\Rightarrow\ K = ((i-1)(i-2))/2 + h\,, \\ h < i \ &\Rightarrow\ K = ((h-1)(h-2))/2 + i\,. \end{aligned} \quad (10.11)$$

The subroutine TIHEXM shown in Table 10.6 expects that the matrix $(t_{ih})$ is passed in the form (10.10). It is naturally unable to check whether the user has done this. It does, however, test whether M12 < (m(m−1))/2, in which case TIHEXM returns immediately with IFLAG = 11.

The choice of method is governed by the parameter METHOD which can take the values 1, 2 and 3, corresponding to the objective functions $M_1$, $M_2$ and $M_3$ defined by (7.7), (7.8) and (7.9). The exchange method is applied to the given starting partition (Z(I), I=1, M), provided that the partition is feasible. The minimum number of elements in a cluster, M0, does not appear as a parameter, but is set inside TIHEXM. For METHOD = 1, 2, M0 = 1, and for METHOD = 3, M0 = 2. This indicates that, for $M_3$, the version shown in (7.10) has been implemented. The other version of $M_3$, shown in (7.11), could easily be included with a few extra statements.

TIHEXM returns IFLAG = 10 when it is called with METHOD not equal to 1, 2 or 3. Further parameters of TIHEXM have the same meaning as for TRWEXM. C is workspace.

Since the distance matrix for m objects requires (m(m−1))/2 storage locations, a fairly large amount of storage is required for big values of m. A situation such as Case 5 in the introduction, where we had m ≈ 5000, reaches the limits of what is currently possible. If we do not want to restrict ourselves to a smaller subset of objects, we can avoid the need to store the values (T(K), K=1, M12), provided that they have not been given directly but have been calculated from a data matrix. In that case, the declaration T(M12) can be omitted from the

*Table 10.6*

```
      SUBROUTINE TIHEXM (M,M12,T,N,MJ,METHOD,Z,E,D,IT,IFLAG,C)
      INTEGER Z(M),MJ(N),P,Q,H
      DIMENSION T(M12),E(N),C(N)
      IT=0
      D=0.
      DO 1 J=1,N
            MJ(J)=0
            E(J)=0.
    1 CONTINUE
      IFLAG=10
      IF(METHOD.LT.1.OR.METHOD.GT.3) GOTO 21
      IFLAG=11
      IF(M12.LT.(M*(M-1))/2) GOTO 21
      IFLAG=0
      BIG=1.E50
      R=.999
      MO=1
      IF(METHOD.EQ.3) MO=2
      DO 3 I=1,M
            J=Z(I)
            IF(J.GE.1.AND.J.LE.N) GOTO 2
            IFLAG=1
            GOTO 21
    2       MJ(J)=MJ(J)+1
    3 CONTINUE
      DO 4 J=1,N
            IF(MJ(J).GE.MO) GOTO 4
            IFLAG=2
            GOTO 21
    4 CONTINUE
      K=0
      DO 6 I=2,M
            P=Z(I)
            L=I-1
            DO 5 H=1,L
                  Q=Z(H)
                  K=K+1
                  IF(Q.EQ.P) E(Q)=E(Q)+T(K)
    5       CONTINUE
    6 CONTINUE
      DO 8 J=1,N
            F=E(J)
            IF(METHOD.EQ.1) GOTO 7
            V=MJ(J)
            IF(METHOD.EQ.2) F=F/V
            IF(METHOD.EQ.3) F=F/(V*(V-1.))
            E(J)=F
    7       D=D+F
    8 CONTINUE
      IF(N.LE.1) GOTO 21
      IS=0
      I=0
    9 IS=IS+1
      IF(IS.GT.M) GOTO 21
   10 I=I+1
      IF(I.LE.M) GOTO 11
      IT=IT+1
      IF(IT.GT.15) GOTO 21
      I=1
   11 P=Z(I)
      L=MJ(P)
      IF(L.LE.MO) GOTO 9
      V=L
      V1=V-1.
```

*Table 10.6 – continued*

```
      IM=((I-1)*(I-2))/2
      DO 12 J=1,N
         C(J)=0.
   12 CONTINUE
      DO 14 H=1,M
         IF(H.EQ.I) GOTO 14
         J=Z(H)
         KH=H
         KI=IM
         IF(I.GT.H) GOTO 13
         KI=I
         KH=((H-1)*(H-2))/2
   13    K=KH+KI
         C(J)=C(J)+T(K)
   14 CONTINUE
      EQ=BIG
      DO 17 J=1,N
         BJ=E(J)
         CJ=C(J)
         U=MJ(J)
         U1=U+1.
         IF(J.NE.P) GOTO 15
         IF(METHOD.EQ.1) EP=CJ
         IF(METHOD.EQ.2) EP=(CJ-BJ)/V1
         IF(METHOD.EQ.3) EP=(CJ-2.*V1*BJ)/(V1*(V-2.))
         GOTO 17
   15    IF(METHOD.EQ.1) EJ=CJ
         IF(METHOD.EQ.2) EJ=(CJ-BJ)/U1
         IF(METHOD.EQ.3) EJ=(CJ-2.*U*BJ)/(U*U1)
   16    IF(EJ.GE.EQ) GOTO 17
         EQ=EJ
         Q=J
         W=U
         W1=U1
   17 CONTINUE
      IF(EQ.GE.EP*R) GOTO 9
      IS=0
      D=D-EP+EQ
      IF(METHOD.GT.1) GOTO 18
      E(P)=E(P)-C(P)
      E(Q)=E(Q)+C(Q)
      GOTO 20
   18 IF(METHOD.GT.2) GOTO 19
      E(P)=(V*E(P)-C(P))/V1
      E(Q)=(W*E(Q)+C(Q))/W1
      GOTO 20
   19 E(P)=(V*V1*E(P)-C(P))/(V1*(V-2.))
      E(Q)=(W*(W-1.)*E(Q)+C(Q))/(W*W1)
   20 MJ(P)=L-1
      MJ(Q)=MJ(Q)+1
      Z(I)=Q
      GOTO 10
   21 RETURN
      END
```

DIMENSION statement, T can be omitted from the parameter list for TIHEXM, and a

FUNCTION T(K),

can be defined. Each time it is called with a given K, the function would find the corresponding object numbers i and h and compute an appropriate distance value from rows i and h of the data matrix. If one wants to leave the expressions

T(K) unchanged in order to avoid making further alterations to TIHEXM, then the FUNCTION must take no further arguments such as the data matrix and the number of rows and columns. In this case, the extra parameters would have to be passed in a COMMON statement in TIHEXM as well as in the FUNCTION. The parameter list in TIHEXM would have to be adapted accordingly. As is shown in a programming example in [B42], this approach requires only ms storage locations for the data matrix, instead of $(m(m-1))/2 + ms$ when a distance matrix is used. The penalty is a much longer computation time.

If TIHEXM is to be modified for unsymmetrical distance matrices, it is first necessary to formulate the objective functions somewhat differently, for example

$$M_1(C) = \sum_{j=1}^{n} \Big( \sum_{i \in C_j} \sum_{h \in C_j} t_{ih} \Big),$$

The exchange formula also need to be modified. Since there is now no point in storing a lower triangular matrix as a one-dimensional array, the implementation is simpler than TIHEXM. On the other hand, twice the storage is required.

## 10.5 CLREXM: The exchange method for clusterwise linear regression using Givens rotations

This section deals with the approximate solution of the problem of clusterwise linear regression analysis using the exchange method with objective function (8.9), and taking account of the extra condition $C \in P'(n, M)$ as defined in (8.10).

The routines to perform the QR-decomposition for feasible initial partitions, and to do the up- and downdating, are based on the ALGOL procedure in [B20] which use Givens rotations. The procedures have been adapted and combined for the present purpose. For details, see the original literature [B10, B19, B20].

Subroutine INEXCL, shown in Table 10.7, does not explicitly calculate or update the regression coefficients ((X(J, K), K=1, S), J=1, N) of the individual regression problems. Instead, it computes the auxiliary arrays T and R used to determine the X(J, K) and the error sums of squares (E(J), J=1, N) according to (8.15), which are used to decide if an exchange takes place.

Subroutine CLREXM implements the exchange method, and is listed in Table 10.8. First, the subroutine checks if the initial partition (Z(I), I=1, M) satisfies the necessary condition $m_j \geqslant s$ $(j=1, \ldots, n)$. The test is governed by the parameter $M0 \geqslant S$. If $M0 < S$, CLREXM returns immediately with IFLAG = 14. The real arrays (A(I, K), K=1, S), I=1, M) and (B(I), I=1, M) are new parameters which contain the values of the independent, and of the dependent, variables. Also new is the parameter NS used to dimension various arrays. It must be set equal to $(s(s-1))/2$ in the calling program. When the parameter is passed with a value smaller than this, CLREXM returns immediately with IFLAG = 12. Numerous workspace arrays AI, ..., RQ are required. Their sizes

*Table 10.7*

```
      SUBROUTINE INEXCL (SDX,S,AI,BI,WI,J,NDX,N,NS,
     *                   F,T,R,FJ,TJ,RJ,SS,NNMAX,KMAX)
      INTEGER SDX,S
      DIMENSION F(SDX,N),T(SDX,N),R(NDX,NS),AI(S),
     *          FJ(S),TJ(S),RJ(NS)
      ZERO=0.
      ONE=1.
      DO 3 K=1,S
            IF(WI.EQ.ZERO) GOTO 4
            AK=AI(K)
            IF(AK.EQ.ZERO) GOTO 3
            FK=F(K,J)
            WA=WI*AK
            DP=FK+WA*AK
            HK=ONE/DP
            FH=FK*HK
            WH=WA*HK
            WI=WI*FH
            FJ(K)=DP
            IF(K.EQ.S) GOTO 2
            K1=K+1
            NN=((K-1)*(S+S-K))/2+1
            DO 1 L=K1,S
                  AL=AI(L)
                  RL=R(J,NN)
                  AI(L)=AL-AK*RL
                  RJ(NN)=FH*RL+WH*AL
                  NNMAX=NN
                  NN=NN+1
1           CONTINUE
2           AL=BI
            TK=T(K,J)
            BI=AL-AK*TK
            TJ(K)=FH*TK+WH*AL
            KMAX=K
3     CONTINUE
      SS=SS+WI*BI*BI
4     RETURN
      END
```

can be seen in the DIMENSION statements. All other parameters, except for X, have the usual meaning.

Unlike our normal practice, we have refrained from using a modified test corresponding to (9.19) for making an exchange, of the form

$$e(C_q \cup \{i\}) - e(C_q) < R\,(e(C_p) - e(C_p \setminus \{i\})),\ 0<R<1.$$

This modification could easily be implemented by adding a statement RR = .999 at the beginning and by replacing the second statement following statement number 18 by

IF(EQ.GE.EP*RR) GOTO 11

Probably, it would be appropriate to reduce the value of RR to less than .999.

The desired regression coefficients ((X(J, K), K=1, S), J=1, N) are calculated after the end of the exchange method and shortly before leaving CLREXM, based on the updated auxiliary arrays T and R [B20].

Since the downdating is in principle ill-conditioned [B53], there is a danger that it will lead to inaccurate or even wrong results when $m_j$ is not distinctly greater than s. In that situation, even numerically very stable methods do not necessarily produce good results. We therefore recommend setting $m_j \approx 2s$, that is, $M0 \approx 2s$. This should be possible provided that the number of classes, n, is sufficiently small.

*Table 10.8*

```
      SUBROUTINE CLREXM (A,MDX,M,SDX,S,B,NDX,N,NS,Z,MO,MJ,X,
     *                   E,D,IT,IFLAG,AI,F,T,R,FJ,TJ,RJ,
     *                   FP,TP,RP,FQ,TQ,RQ)
      INTEGER SDX,S,Z(M),MJ(N),P,Q,S1,S2
      DIMENSION A(MDX,S),B(M),X(NDX,S),E(N)
      DIMENSION F(SDX,N),T(SDX,N),R(NDX,NS),FJ(S),TJ(S),RJ(NS),
     *          FP(S),TP(S),RP(NS),FQ(S),TQ(S),RQ(NS),AI(S)
      IT=0
      ZERO=0.
      ONE=1.
      BIG=1.E50
      D=ZERO
      IFLAG=14
      IF(MO.LT.S) GOTO 26
      IFLAG=12
      IF(NS.LT.((S-1)*S)/2) GOTO 26
      IFLAG=0
      S1=S+1
      S2=S+S
      DO 3 J=1,N
            MJ(J)=0
            E(J)=ZERO
            DO 1 K=1,S
                  F(K,J)=ZERO
                  T(K,J)=ZERO
1           CONTINUE
            DO 2 NN=1,NS
                  R(J,NN)=ZERO
2           CONTINUE
3     CONTINUE
      DO 8 I=1,M
            J=Z(I)
            IF(J.GE.1.AND.J.LE.N) GOTO 4
            IFLAG=1
            GOTO 26
4           MJ(J)=MJ(J)+1
            WI=ONE
            BI=B(I)
            DO 5 K=1,S
                  AI(K)=A(I,K)
5           CONTINUE
            CALL INEXCL (SDX,S,AI,BI,WI,J,NDX,N,NS,F,T,R,
     *                   FJ,TJ,RJ,E(J),NNMAX,KMAX)
            DO 6 K=1,KMAX
                  F(K,J)=FJ(K)
                  T(K,J)=TJ(K)
6           CONTINUE
            DO 7 NN=1,NNMAX
                  R(J,NN)=RJ(NN)
7           CONTINUE
8     CONTINUE
      DO 10 J=1,N
            IF(MJ(J).GE.MO) GOTO 9
            IFLAG=2
            GOTO 26
9           D=D+E(J)
```

*Table 10.8 – continued*

```
10 CONTINUE
   IF(N.LE.1) GOTO 21
   IS=0
   I=0
11 IS=IS+1
   IF(IS.GT.M) GOTO 21
12 I=I+1
   IF(I.LE.M) GOTO 13
   IT=IT+1
   IF(IT.GT.15) GOTO 21
   I=1
13 P=Z(I)
   IF(MJ(P).LE.MO) GOTO 11
   EQ=BIG
   DO 18 J=1,N
         EJ=E(J)
         BI=B(I)
         DO 14 K=1,S
               AI(K)=A(I,K)
14       CONTINUE
         WI=ONE
         IF(J.NE.P) GOTO 15
         EP=EJ
         CALL INEXCL (SDX,S,AI,BI,-WI,P,NDX,N,NS,F,T,R,
  *                       FP,TP,RP,EP,NNMAX,KMAX)
         CP=EP
         GOTO 18
15       CALL INEXCL (SDX,S,AI,BI,WI,J,NDX,N,NS,F,T,R,
  *                       FJ,TJ,RJ,EJ,NNMAX,KMAX)
         CJ=EJ
         EJ=CJ-E(J)
         IF(EJ.GE.EQ) GOTO 18
         EQ=EJ
         CQ=CJ
         Q=J
         DO 16 K=1,S
               FQ(K)=FJ(K)
               TQ(K)=TJ(K)
16       CONTINUE
         DO 17 NN=1,NS
               RQ(NN)=RJ(NN)
17       CONTINUE
18 CONTINUE
   EP=E(P)-CP
   IF(EQ.GE.EP) GOTO 11
   IS=0
   MJ(P)=MJ(P)-1
   MJ(Q)=MJ(Q)+1
   D=D-EP+EQ
   E(P)=CP
   E(Q)=CQ
   DO 19 K=1,S
         F(K,P)=FP(K)
         F(K,Q)=FQ(K)
         T(K,P)=TP(K)
         T(K,Q)=TQ(K)
19 CONTINUE
   DO 20 NN=1,NS
         R(P,NN)=RP(NN)
         R(Q,NN)=RQ(NN)
20 CONTINUE
   Z(I)=Q
   GOTO 12
21 DO 25 J=1,N
         DO 24 KK=1,S
               K=S1-KK
               TJ(K)=T(K,J)
               IF(K.EQ.S) GOTO 23
```

*Table 10.8 – continued*

```
                NN=((K-1)*(S2-K))/2+1
                K1=K+1
                DO 22 L=K1,S
                     TJ(K)=TJ(K)-R(J,NN)*TJ(L)
                     NN=NN+1
22              CONTINUE
23              X(J,K)=TJ(K)
24         CONTINUE
25 CONTINUE
26 RETURN
   END
```

The author has found the use of double precision (which gives 24 decimal places on the TR 440 computer) to be very successful, although it is no panacea. This requires only a few changes in INEXCL, CLREXM, and in the calling program. Where an IBM compiler is used, an additional first statement

IMPLICIT REAL*8 (A – H, O – Z)

needs to be added. The DIMENSION declarations need to be changed to REAL*8, and in INEXCL and CLREXM the two statements

ZERO = 0.
ONE = 1.

need to be changed into

ZERO = 0.D0
ONE = 1.D0

and finally, in CLREXM

BIG = 1.E50

needs to be replaced by

BIG = 1.D50

Since INEXCL uses the form of the Givens rotation which does not use square roots [B19], no further changes are required.

In [B50] a different version of CLREXM is shown in which parameters are passed via COMMON statements. In general, this leads to faster object code.

**Problem 10.1**: Implement the minimal distance method for the $L_1$ criterion, where the data matrix contains only integers. Are different versions, corresponding to OVSEXM and OVREXM for the exchange method, either possible or useful?

**Problem 10.2**: Change the subroutine OVPEXM as indicated in Section 10.2, so that it can cope with ordinal vectors whose individual components have different ranges $1, \ldots, t_k$ $(k=1, \ldots, s)$.

**Problem 10.3:** Write a subroutine which performs the transformation of an ordinal into a binary data matrix, as defined in (10.6). Declare the binary matrix as LOGICAL so that it can be processed by BVPEXM.

**Problem 10.4:** Rewrite the inverse of the storage mapping (10.10) in a form which allows it to be programmed.

**Problem 10.5:** Write a subroutine to calculate the Jaccard metric defined in (7.3) for binary or ordinal vectors which are grouped together in a data matrix X, and store the results according to (10.10) so that you can apply TIHEXM directly.

**Problem 10.6:** Write an alternative implementation of the minimal distance method for (8.9), using INEXCL or HFTI in [B29].

# Part III

# Simple main programs, examples, suggestions for use

# 11 Sample main programs, examples, and evaluation for TRWEXM, TRWMDM, DETEXM, and DWBEXM

## 11.1 Pseudo-random generation of initial partitions

All the subroutines described, TRWMDM, TRWEXM, DETEXM, DWBEXM, OVSEXM, OVREXM, OVPEXM, BVPEXM, TIHEXM and CLREXM require an initial partition. It does not matter how this is generated. However, if the initial partition, as described at the end of Section 9.1, is not feasible for the particular method employed, all subroutines will immediately return with IFLAG = 1 or IFLAG = 2. When the initial partition is feasible, the minimal distance or exchange method is set in motion. With the exchange method, the minimum number of elements in each cluster is maintained explicitly via the parameter M0. Using the minimal distance method, this minimum may not be maintained, in which case IFLAG = 2 is returned. Other constraints are $C \in P^{+}(n,M)$ for DETEXM, $C \in P^{++}(n,M)$ for DWBEXM and $C \in P'(n,M)$ for CLREXM. Whether these hold either for starting partitions or during an exchange attempt can only be determined retrospectively in a numerical sense.

To gain an impression about the behaviour of the procedures using different objective functions, our main testing programs will use NRMAX (generally 20) pseudo-random initial partitions. These are generated by subroutine RANDP, shown in Table 11.1. RANDP itself uses URAND [B17] shown in Table 11.2 and discussed in the next paragraph, which generates uniformly distributed pseudo-random numbers in the interval [0, 1). The numbers generated depend on the seed IY and on the word length of the computer used. Their period is very large. In RANDP, the interval [0, 1) is split into N subintervals of equal size. For I=1, M, a random number PS = URAND(IY) is generated, which falls within subinterval J. Z(I) is then set equal to J. This method avoids the error IFLAG = 1. However, if the partition generated does not have at least M0 elements with Z(I) = J for J=1, M, then error IFLAG = 2 will occur. When M is sufficiently large compared with N, usually this does not happen.

The use of URAND assumes that, on the computer used, integers are represented as binary numbers, and that the multiplication of two integers is performed modulo a power of two [B17]. If this is not the case, a different generator must be used (see Appendix). When URAND is first called, it finds out the word length of integers on the computer and uses this to ensure that the period of the pseudo-random numbers generated is as large as possible. When it is first called, URAND must be given a seed IY which can be any integer. Each execution of a statement of the form PS = URAND(IY) generates another uniformly distributed pseudo-random number, $0 \leqslant PS < 1$, and changes IY, ready for the next call to URAND (which may, of course, be done within an arithmetic expression). Because of the dependence on word length, different computers will in general produce different sequences of pseudo-random numbers, even when the seed IY is the same.

*Table 11.1*

```
      SUBROUTINE RANDP (M,N,Z,IY)
      INTEGER Z(M),ZI
      DO 1 I=1,M
           Z(I)=1
    1 CONTINUE
      IF(N.LE.1.OR.M.LE.1) GOTO 5
      H=1./N
      DO 4 I=1,M
           PS=URAND(IY)
           F=H
           ZI=1
    2      IF(ZI.EQ.N) GOTO 3
           IF(PS.LT.F) GOTO 3
           F=F+H
           ZI=ZI+1
           GOTO 2
    3      Z(I)=ZI
    4 CONTINUE
    5 RETURN
      END
```

*Table 11.2*

```
      FUNCTION URAND(IY)
      INTEGER  IY
      INTEGER  IA,IC,ITWO,M2,M,MIC
      DOUBLE PRECISION  HALFM
      REAL  S
      DOUBLE PRECISION  DATAN,DSQRT
      DATA M2/0/,ITWO/2/
      IF (M2 .NE. 0) GO TO 20
      M = 1
   10 M2 = M
      M = ITWO*M2
      IF (M .GT. M2) GO TO 10
      HALFM = M2
      IA = 8*IDINT(HALFM*DATAN(1.D0)/8.D0) + 5
      IC = 2*IDINT(HALFM*(0.5D0-DSQRT(3.D0)/6.D0)) + 1
      MIC = (M2 - IC) + M2
      S = 0.5/HALFM
   20 IY = IY*IA
      IF (IY .GT. MIC) IY = (IY - M2) - M2
      IY = IY + IC
      IF (IY/2 .GT. M2) IY = (IY - M2) - M2
      IF (IY .LT. 0) IY = (IY + M2) + M2
      URAND = FLOAT(IY)*S
      RETURN
      END
```

Consequently, RANDP may produce different initial partitions on different computers. This implies that the results of our test examples for NRMAX starting partitions cannot necessarily be reproduced exactly using the main programs shown. This may sound unfortunate but cannot be avoided, since the author knows of no pseudo-random number generator in FORTRAN which is guaranteed to produce the same sequence of numbers on every computer. However, since NRMAX initial partitions give us a whole distribution of values of the objective functions corresponding to end partitions, it can be assumed that the distributions turn out very similar. A frequently occurring minimum will also occur frequently on other computers.

## 11.2 Examples for the variance criterion and their display

We shall illustrate the variance criterion and compare the minimal distance and exchange methods and their combination (see the remark immediately before Theorem 2.8). As test examples we choose four data matrices ((X(I, K), K=1, 2), I=1, M) with different numbers of rows M, but all with S = 2 columns. The

*Table 11.3*

| | | | | | | | | | | | | | | | |
|---|---|---|---|---|---|---|---|---|---|---|---|---|---|---|---|
| 41 | 39 | 42 | 44 | 10 | 38 | 8 | 41 | 13 | 45 | 7 | 38 | 42 | 9 | 12 | 19 |
| 25 | 6 | 13 | 9 | 12 | 32 | 26 | 39 | 34 | 37 | 22 | 38 | 35 | 31 | 26 | 38 |
| 29 | 34 | 37 | 40 | 42 | | | | | | | | | | | |
| | | | | | | | | | | | | | | | |
| 45 | 44 | 43 | 43 | 42 | 42 | 41 | 41 | 40 | 40 | 39 | 39 | 39 | 38 | 38 | 38 |
| 38 | 37 | 35 | 34 | 34 | 27 | 25 | 24 | 23 | 23 | 22 | 21 | 20 | 18 | 16 | 13 |
| 11 | 11 | 10 | 9 | 9 | | | | | | | | | | | |
| | | | | | | | | | | | | | | | |
| 13 | 41 | 50 | 6 | 22 | 33 | 10 | 18 | 24 | 43 | 37 | 12 | 19 | 28 | 35 | 1 |
| 9 | 50 | 41 | 25 | 19 | 6 | 33 | 43 | 17 | 25 | 8 | 38 | 14 | 50 | 1 | 24 |
| 44 | 34 | 42 | 7 | 17 | 35 | 26 | 48 | 5 | | | | | | | |
| | | | | | | | | | | | | | | | |
| 50 | 48 | 46 | 45 | 45 | 45 | 41 | 40 | 39 | 37 | 36 | 34 | 33 | 33 | 32 | 31 |
| 31 | 30 | 28 | 27 | 25 | 24 | 23 | 22 | 20 | 19 | 17 | 17 | 16 | 16 | 14 | 14 |
| 12 | 11 | 7 | 6 | 6 | 5 | 2 | 2 | 1 | | | | | | | |
| | | | | | | | | | | | | | | | |
| 38 | 40 | 44 | 46 | 49 | 42 | 45 | 47 | 14 | 44 | 48 | 50 | 29 | 43 | 46 | 49 |
| 7 | 41 | 48 | 50 | 45 | 38 | 42 | 47 | 50 | 44 | 48 | 33 | 47 | 50 | 42 | 10 |
| 46 | 21 | 2 | 34 | 42 | 48 | 22 | 36 | 7 | 26 | 41 | 17 | | | | |
| | | | | | | | | | | | | | | | |
| 50 | 50 | 50 | 50 | 50 | 49 | 49 | 49 | 48 | 48 | 48 | 48 | 47 | 47 | 47 | 47 |
| 45 | 45 | 45 | 45 | 44 | 43 | 43 | 43 | 43 | 41 | 41 | 40 | 40 | 40 | 39 | 38 |
| 38 | 36 | 35 | 34 | 34 | 27 | 26 | 25 | 14 | 14 | 7 | 5 | | | | |
| | | | | | | | | | | | | | | | |
| 24 | 38 | 19 | 23 | 22 | 24 | 26 | 42 | 23 | 40 | 10 | 41 | 44 | 40 | 44 | 20 |
| 3 | 6 | 8 | 26 | 43 | 7 | 13 | 38 | 6 | 8 | 10 | 5 | 26 | 38 | 9 | 31 |
| 35 | 46 | 11 | 28 | 34 | 36 | 18 | 37 | 35 | 39 | 37 | 9 | 41 | 31 | 38 | 23 |
| 50 | 11 | 31 | 39 | 44 | 7 | 21 | 39 | 12 | 16 | 5 | 44 | 9 | 11 | 42 | 10 |
| 13 | 41 | 43 | 9 | 11 | 43 | 10 | 27 | 14 | | | | | | | |
| | | | | | | | | | | | | | | | |
| 49 | 45 | 44 | 44 | 43 | 43 | 43 | 43 | 42 | 42 | 41 | 41 | 41 | 40 | 40 | 39 |
| 38 | 37 | 37 | 37 | 37 | 36 | 36 | 36 | 35 | 35 | 34 | 33 | 33 | 33 | 32 | 30 |
| 30 | 30 | 28 | 28 | 27 | 27 | 26 | 26 | 25 | 25 | 24 | 23 | 23 | 21 | 21 | 19 |
| 19 | 18 | 17 | 16 | 15 | 14 | 13 | 11 | 10 | 9 | 8 | 8 | 7 | 7 | 7 | 6 |
| 6 | 6 | 6 | 5 | 5 | 5 | 3 | 3 | 1 | | | | | | | |

choice of S = 2 has the great advantage that the data points and cluster results can be displayed simply in a two-dimensional Cartesian coordinate system, and are therefore easy to compare with the results of other methods. Because of the geometric interpretation, and because the numbers in the two columns can often be regarded as multiples of the same unit of length, we have no obvious scaling problems. However, we must be aware that end partitions, including optimal ones, could look different if each of the two columns was measured in different units.

In choosing the four examples, we were guided by the geometric configuration and not by assumptions such as the presence of a mixture of normal distributions which are, in any case, not verifiable when $s > 2$. The data matrices for the four examples are shown numerically in Table 11.3 (rows and columns being transposed), and graphically in Fig. 11.1 with the aid of subroutine PRINT, which is listed in Table 11.4.

*Table 11.4*

```
      SUBROUTINE PRINT(M,X1,X2,P,N)
      INTEGER X1(M),X2(M),P(M),Q(9)
      INTEGER BLANK,ZERO
      DATA BLANK/' '/, ZERO/'0'/,
     *     Q/'1','2','3','4','5','6','7','8','9'/
      KO=6
      IF(N.GT.9) GOTO 8
      WRITE(KO,5)
      K=1
      DO 4 I=1,50
           WRITE(KO,7)
           JM=51-I
1          IF(X2(K).NE.JM) GOTO 4
           KM=X1(K)+10
           KP=P(K)
           DO 2 J=1,N
                IF(KP.NE.J) GOTO 2
                KP=Q(J)
                GOTO 3
2          CONTINUE
           KP=ZERO
3          WRITE(KO,6) (BLANK,L=1,KM),KP
           K=K+1
           IF(K.GT.M) GOTO 8
           GOTO 1
4     CONTINUE
5     FORMAT('1')
6     FORMAT('+',132A1)
7     FORMAT(' ')
8     RETURN
      END
```

The first example is a data set with a structure, that is, three or four classes do clearly exist. Provided that the right number of clusters is specified, each objective function ought to find them. The second example is a totally unstructured data set of a kind that occurs quite commonly in practical applications. The third data set is also almost without structure. Only in one place is there an accumulation of objects. The fourth and last data set does contain several accumulations of objects. It is likely to be much more representative than the first set of the situation encountered in practical applications, when a structure is to be discovered.

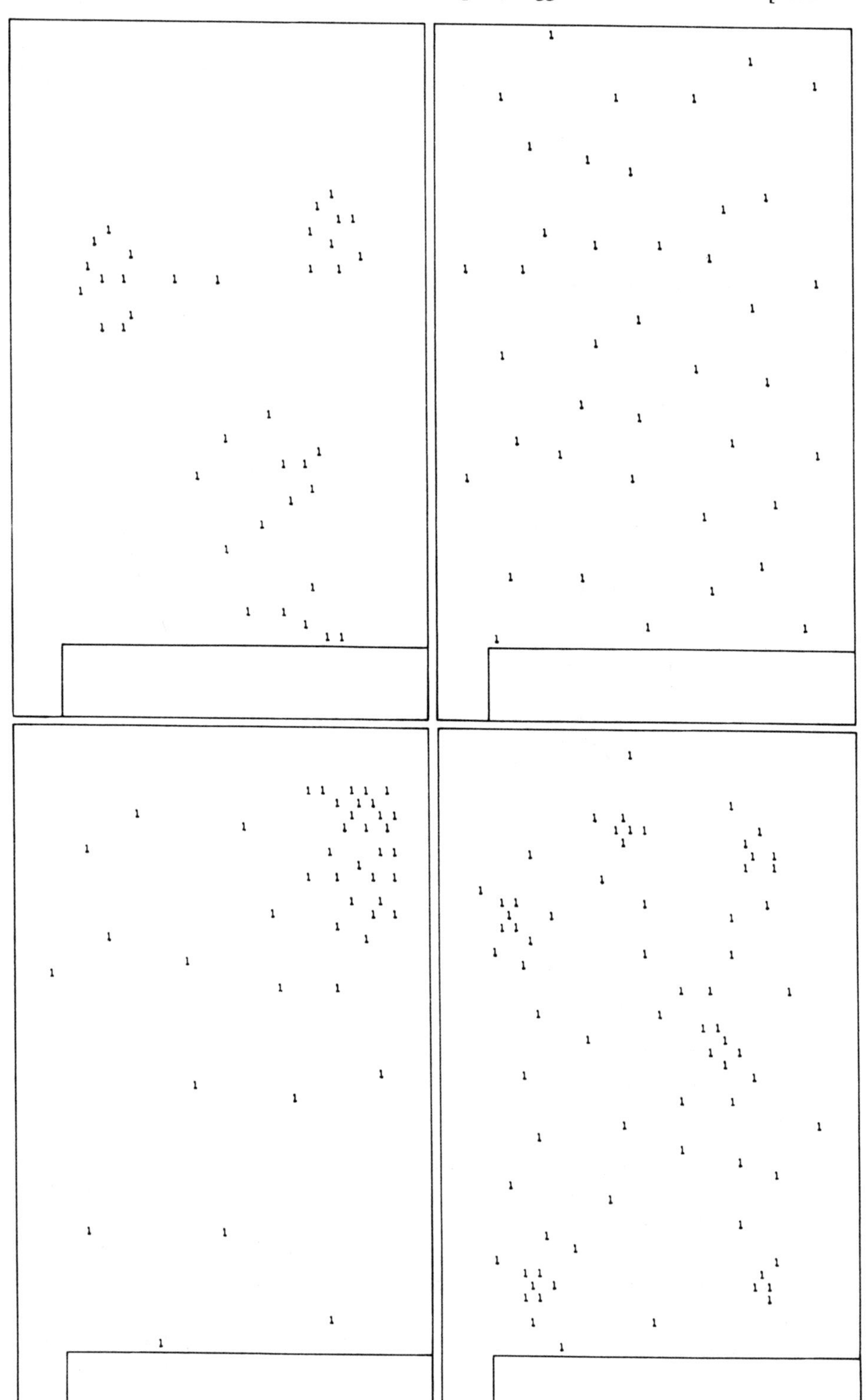

*Fig. 11.1*

In order to be displayed on a printer using subroutine PRINT, the examples must have the following properties. Using the notation (X(I, 1) = X1(I), X(I, 2) = X2(I), I = 1, M) we must have $1 \leqslant$ X1(I) $\leqslant$ M1, $1 \leqslant$ X2(I) $\leqslant$ M2 in these and in four later examples. Here X1(I) and X2(I) must be integers. We must have M1 $\leqslant 122$ (we only use M1 $\leqslant 50$) and M2 $\leqslant 50$. Furthermore, the pairs (X1(I), X2(I)) must be sorted in descending order of the X2(I). For X2(I) = X2(I + 1) it does not matter whether XI(I) is greater than, equal to, or less than X1(I+1). Moreover, the number N of classes must not exceed 9. If N > 9, PRINT returns without indicating an error. If each point is associated with a class number P(I), $1 \leqslant$ P(I) $\leqslant$ N $\leqslant 9$, the value of P(I) will be printed in the coordinate position of point I. The FORTRAN unit number associated with the printer is KO = 6. If P(I) is not in the range from 1 to 9, '0' is printed. Thus, a visual representation of the classes found is obtained. To illustrate the example in Fig. 11.1, P(I) was set equal to 1 for I=1, M. Among the subroutines, PRINT needs to be changed when using a FORTRAN 77 instead of a FORTRAN IV compiler.

Since the width of a character on the printer is less than the line spacing (the ratio tends to be 3:5), an undistorted representation is not possible. In interpreting the Figures it must be borne in mind that on the printed graph X2(I) = $\beta$ X1(I), where $\beta \approx 1.67$. With scale-dependent criteria such as the variance criterion, it could be that end partitions in this scale might differ from those in the 1:1 scale. This does not matter in the case of DETEXM and DWBEXM. However, even there, the distortion cannot simply be ignored when the results are looked at and compared with those of the variance criterion.

## 11.3 Sample main programs for TRWEXM, TRWMDM, and for a combination of the two

The sample main program for TRWEXM is shown in Table 11.5. In a modified form, it is also used with all other subroutines, and it is only described in detail in the current section. It has two aims. First, the exchange method is to be carried out with NRMAX starting partitions generated by RANDP, and the resulting values of the objective function ((DNNR(N, NR), NR=1, NRMAX), N=N1, N2) and the numbers of iterations needed ((KIT(N, NR), NR=1, NRMAX), N=N1, N2) are to be recorded systematically for numbers of classes ranging from N1 = 2 to N2 = 6. If IFLAG = 2, DNNR (N, NR) is set equal to 0., and KIT(N, NR) = – IFLAG. All this happens when KPRINT = 0. When KPRINT = 1, all those end partitions are in addition displayed using PRINT which have a smaller value of the objective function D than the previous minimum DD. Initially, DD is set equal to 1.E50, and before each comparison, multiplication by R = .999 takes place. The output produced with PRINT is annotated with M, N, NR, IT, D and (J, M(CJ), E(J), J=1, N), as well as with an identifier for the objective function used – in this case TRW. It is organized in this way to avoid problems of moving stored data and because it is of interest to see how much difference there is between the worst and the best end partitions. However, to save space when presenting the examples, we have shown the plots only for the best of the NRMAX end partitions

The program uses units KI = 5 for input and KO = 6 for output. The input is read in FORMAT (16I5). On the first card image, the program reads M ($\leqslant$ 200), and on subsequent card images (X1(I), I=1, M) and (X2(I), I=1, M), each starting on a new card. Several such simple sets of input data may follow one after another. (In Table 11.3, M is omitted, and a blank line is inserted between X1 and X2.) The program stops when M $\leqslant$ 0.

Since TRWEXM and TRWMDM have the same parameter list, the main program can be made to handle TRWMDM by altering the subroutine name in CALL TRWEXM (...). In that case, KIT(N, NR) stores the number of minimal distance steps, and this is subsequently printed. When N1 and NRMAX are unchanged, the same starting partitions as before will be used. This is also true for the successive use of the minimal distance and exchange methods. Here, the program in Table 11.5 merely needs the addition of a statement CALL TRWMDM (...) just before the statements call TRWEXM (...). Should TRWMDM have broken off with IFLAG = 2, this will be recognized in TRWEXM when MEANS is called, and another return with IFLAG = 2 is made.

*Table 11.5*

```
C   SAMPLE MAIN PROGRAM FOR TRWMDM, TRWEXM, OR
C   TRWMDM ---> TRWEXM SUCCESSIVELY
C
C   MDX=200       MAXIMAL NUMBER OF OBJECTS
C   S=2           NUMBER OF VARIABLES (REQUIRED FOR GRAPHICAL OUTPUT)
C   NDX=9         MAXIMAL NUMBER OF CLUSTERS
C   MO=1          MINIMAL NUMBER OF CLUSTER MEMBERS
C
      INTEGER   Z(200),X1(200),X2(200),KIT(9,20),MJ(9)
      DIMENSION X(200,2),XBAR(9,2),E(9),DNNR(9,20)
      KI=5
      KO=6
      BIG=1.E50
      R=.999
      KPRINT=1
      MO=1
      N1=2
      N2=6
      NRMAX=20
    1 READ(KI,20) M
      IF(M.LE.0) STOP
      READ(KI,20) (X1(I),I=1,M)
      READ(KI,20) (X2(I),I=1,M)
      DO 2 I=1,M
      X(I,1)=X1(I)
    2 X(I,2)=X2(I)
      DO 6 N=N1,N2
            DD=BIG
            IY=37519
            DO 5 NR=1,NRMAX
                  CALL RANDP (M,N,Z,IY)
                  CALL TRWEXM (X,200,M,2,Z,MO,MJ,XBAR,9,N,
     *                            E,D,IT,IFLAG)
                  IF(IFLAG.EQ.0) GOTO 3
                  DNNR(N,NR)=0.
                  KIT(N,NR)= -IFLAG
                  GOTO 5
    3             DNNR(N,NR)=D
                  KIT(N,NR)=IT
                  IF(D.GE.DD*R) GOTO 5
                  IF(KPRINT.EQ.0) GOTO 4
                  CALL PRINT (M,X1,X2,Z,N)
                  WRITE(KO,22)
                  WRITE(KO,21) M,N,NR,IT,D
                  WRITE(KO,22) (J,MJ(J),E(J),J=1,N)
    4             DD=D
    5       CONTINUE
    6 CONTINUE
      WRITE(KO,23) (J,J=N1,N2)
      DO 7 NR=1,NRMAX
    7 WRITE(KO,24) NR,(KIT(J,NR),DNNR(J,NR),J=N1,N2)
      GOTO 1
   20 FORMAT(16I5)
   21 FORMAT('0',17X,'M=',I3,'  N=',I1,'  NR=',I2,'  IT=',
     *        I2,'  D=',E11.4,'  (TRW)')
   22 FORMAT('0',17X,2(I1,' | (',I2,')  ',E11.4,5X)/
     *            (18X,2(I1,' | (',I2,')  ',E11.4,5X)))
   23 FORMAT('1',17X,I1,8I14)
   24 FORMAT(' ',I5,9(I3,E11.4))
      END
```

### 11.4 Comparison of the minimal distance and exchange methods, and of both in combination

We have run the main program listed in Table 11.5 both in its original form and with the modifications described at the end of the previous section, using the four sample sets of data introduced in Section 11.2. The results are recorded in four tables, 11.6 to 11.9, showing the values of the objective function and the number of passes or minimal distance steps for NRMAX = 20 random starting partitions and for N = 2, . . ., 6 clusters. The top third of each table shows results for TRWEXM, the centre those for TRWMDM, and the bottom those for a combination of both methods. The starting partitions were the same in each case. Note the values of the objective function, such as .1109E5, .1111E5, and .1128E5 in Table 11.7, do not belong to the same end partition eventually because of rounding errors. They do belong to distinct, but not very different, end partitions. It is not our intention to make authoritative statements about the performance of the three methods, based on only four examples, nor do we wish to give that impression. It would indeed be interesting and important to make general theoretical statements about the complexity of the problem, indicating how good are the values of the objective function achievable by the various methods. However, no such statements are known even about worst cases, nor any of a statistical nature. The following observations can only be indicative. The reader should verify the observations using the tables, add to them, and do measurements on examples of his own.

With the exchange method (top section of the tables), the number of passes lies between 1 and 9 for each example, independent of the number of clusters. Usually, five passes suffice. (Since the author knows of no example, even with $s > 2$, where more than 15 passes are needed, the algorithm is limited to 15 passes, as discussed earlier.) The number of different values of the objective function increases as the number of clusters increases, suggesting that NRMAX might be made larger with increasing N. Where there is a natural structure, as in example 1 for $n = 2, 3, 4, 5$, in example 3 for $n = 2, 3$ and in example 4 for $n = 3, 4$ (see also Fig. 11.2 to 11.6 in the next section), this manifests itself by getting the same value of the objective function for almost all starting partitions. Presumably this is the optimum value.

With the minimal distance method (centre sections) the number of minimal distance iterations usually increases with the number of classes requested. The number of cases in which fewer classes are produced than requested (IFLAG = 2) also tends to increase. This does not appear to depend on the ratio m:n (see examples 2 and 4), but seems to happen more often when a larger number of classes is requested than that which exists 'naturally' (see examples 1 and 4 for $n \geqslant 5$). Finally, together with an increase in the number of classes there is an increase in the number of different values of the objective function obtained, as is the case with the exchange method. However, the variance of these values also increases, which does not happen with the exchange method.

*Table 11.6*

| | | 2 | | 3 | | 4 | | 5 | | 6 |
|---|---|---|---|---|---|---|---|---|---|---|
| 1 | 2 | 0,5230E+04 | 2 | 0,1558E+04 | 2 | 0,9888E+03 | 2 | 0,7502E+03 | 5 | 0.5951E+03 |
| 2 | 2 | 0,5230E+04 | 2 | 0,1558E+04 | 2 | 0,9888E+03 | 2 | 0,7502E+03 | 4 | 0.5061E+03 |
| 3 | 2 | 0,5230E+04 | 2 | 0,1558E+04 | 1 | 0,9888E+03 | 2 | 0,7502E+03 | 3 | 0.5951E+03 |
| 4 | 2 | 0,5230E+04 | 2 | 0,1558E+04 | 2 | 0,9888E+03 | 4 | 0,7502E+03 | 4 | 0.5061E+03 |
| 5 | 2 | 0,5230E+04 | 2 | 0,1558E+04 | 2 | 0,9888E+03 | 2 | 0,7502E+03 | 2 | 0.5951E+03 |
| 6 | 2 | 0,5230E+04 | 2 | 0,1558E+04 | 2 | 0,9888E+03 | 2 | 0,7502E+03 | 5 | 0.5951E+03 |
| 7 | 2 | 0,5230E+04 | 2 | 0,1558E+04 | 2 | 0,9888E+03 | 3 | 0,7502E+03 | 2 | 0.5951E+03 |
| 8 | 1 | 0,6021E+04 | 2 | 0,1558E+04 | 4 | 0,9888E+03 | 2 | 0,7502E+03 | 2 | 0.5951E+03 |
| 9 | 2 | 0,5230E+04 | 2 | 0,1558E+04 | 2 | 0,9888E+03 | 2 | 0,7502E+03 | 4 | 0.5951E+03 |
| 10 | 2 | 0,5230E+04 | 2 | 0,1558E+04 | 4 | 0,9888E+03 | 2 | 0,7502E+03 | 2 | 0.5951E+03 |
| 11 | 1 | 0,5230E+04 | 2 | 0,1558E+04 | 1 | 0,1138E+04 | 2 | 0,7502E+03 | 2 | 0.5061E+03 |
| 12 | 2 | 0,5230E+04 | 3 | 0,1558E+04 | 1 | 0,9888E+03 | 4 | 0,7502E+03 | 2 | 0.5061E+03 |
| 13 | 3 | 0,5230E+04 | 2 | 0,1558E+04 | 4 | 0,9888E+03 | 3 | 0,7502E+03 | 3 | 0.5061E+03 |
| 14 | 2 | 0,5230E+04 | 2 | 0,1558E+04 | 2 | 0,9888E+03 | 2 | 0,7502E+03 | 3 | 0.5951E+03 |
| 15 | 2 | 0,5230E+04 | 2 | 0,1558E+04 | 2 | 0,9888E+03 | 2 | 0,7502E+03 | 4 | 0.5061E+03 |
| 16 | 2 | 0,5230E+04 | 2 | 0,1558E+04 | 4 | 0,9888E+03 | 2 | 0,7502E+03 | 4 | 0.5951E+03 |
| 17 | 2 | 0,5230E+04 | 2 | 0,1558E+04 | 6 | 0,9888E+03 | 2 | 0,7502E+03 | 5 | 0.5951E+03 |
| 18 | 2 | 0,5230E+04 | 2 | 0,1558E+04 | 2 | 0,9888E+03 | 3 | 0,7502E+03 | 2 | 0.5951E+03 |
| 19 | 2 | 0,5230E+04 | 2 | 0,1558E+04 | 1 | 0,9888E+03 | 2 | 0,7502E+03 | 4 | 0.5061E+03 |
| 20 | 2 | 0,6021E+04 | 2 | 0,1558E+04 | 1 | 0,9888E+03 | 4 | 0,7502E+03 | 4 | 0.5951E+03 |
| | | | | | | | | | | |
| | | 2 | | 3 | | 4 | | 5 | | 6 |
| 1 | 0 | 0.5230E+04 | 0 | 0,1558E+04 | 3 | 0.9888E+03 | -2 | 0.0 | 1 | 0.5061E+03 |
| 2 | 0 | 0.5230E+04 | 0 | 0,1558E+04 | 0 | 0.1314E+04 | 3 | 0.7447E+03 | -2 | 0.0 |
| 3 | 0 | 0.5230E+04 | 1 | 0,1558E+04 | 0 | 0.1314E+04 | -2 | 0.0 | -2 | 0.0 |
| 4 | 0 | 0.5230E+04 | 0 | 0,1558E+04 | 3 | 0.9888E+03 | 0 | 0.7502E+03 | 2 | 0.5061E+03 |
| 5 | 0 | 0.5230E+04 | 0 | 0,1558E+04 | -2 | 0.0 | 3 | 0.7447E+03 | -2 | 0.0 |
| 6 | 0 | 0.5230E+04 | 0 | 0,1558E+04 | 3 | 0.9888E+03 | 0 | 0.7502E+03 | -2 | 0.0 |
| 7 | 0 | 0.7627E+04 | 0 | 0,1558E+04 | -2 | 0.0 | 3 | 0.7447E+03 | 1 | 0.5061E+03 |
| 8 | 0 | 0.6021E+04 | 0 | 0,1558E+04 | 3 | 0.9888E+03 | -2 | 0.0 | -2 | 0.0 |
| 9 | 0 | 0.5230E+04 | 0 | 0,1558E+04 | 3 | 0.9888E+03 | -2 | 0.0 | -2 | 0.0 |
| 10 | 0 | 0.5230E+04 | 0 | 0,1558E+04 | 3 | 0.9888E+03 | 0 | 0.7447E+03 | -2 | 0.0 |
| 11 | 0 | 0.5230E+04 | 0 | 0,1558E+04 | 3 | 0.9888E+03 | 3 | 0.7447E+03 | -2 | 0.0 |
| 12 | 0 | 0.5230E+04 | 0 | 0,1558E+04 | 0 | 0.1314E+04 | -2 | 0.0 | -2 | 0.0 |
| 13 | 0 | 0.5230E+04 | 0 | 0,1558E+04 | 0 | 0.1138E+04 | 3 | 0.7447E+03 | 1 | 0.9212E+03 |
| 14 | 0 | 0.5230E+04 | 0 | 0,1558E+04 | 2 | 0.9888E+03 | -2 | 0.0 | 2 | 0.5061E+03 |
| 15 | 0 | 0.5230E+04 | 0 | 0,1558E+04 | 3 | 0.9888E+03 | 1 | 0.7502E+03 | -2 | 0.0 |
| 16 | 0 | 0.6021E+04 | 0 | 0,1558E+04 | 0 | 0.1314E+04 | -2 | 0.0 | -2 | 0.0 |
| 17 | 0 | 0.5230E+04 | 0 | 0,1558E+04 | -2 | 0.0 | -2 | 0.0 | -2 | 0.0 |
| 18 | 0 | 0.6021E+04 | 0 | 0,1558E+04 | 2 | 0.9888E+03 | -2 | 0.0 | -2 | 0.0 |
| 19 | 0 | 0.5230E+04 | 0 | 0,1558E+04 | 3 | 0.9888E+03 | 1 | 0.7764E+03 | -2 | 0.0 |
| 20 | 0 | 0.6021E+04 | 0 | 0,1558E+04 | 3 | 0.9888E+03 | -2 | 0.0 | 1 | 0.5061E+03 |
| | | | | | | | | | | |
| | | 2 | | 3 | | 4 | | 5 | | 6 |
| 1 | 5 | 0.5230E+04 | 4 | 0,1558E+04 | 5 | 0.1136E+04 | -2 | 0.0 | 10 | 0.5107E+03 |
| 2 | 5 | 0.5230E+04 | 6 | 0,1558E+04 | 5 | 0.1314E+04 | 4 | 0.9698E+03 | -2 | 0.0 |
| 3 | 5 | 0.5230E+04 | 5 | 0.4986E+04 | 4 | 0.1314E+04 | -2 | 0.0 | -2 | 0.0 |
| 4 | 4 | 0.5230E+04 | 3 | 0,1558E+04 | 5 | 0.1136E+04 | 6 | 0.7502E+03 | 6 | 0.6156E+03 |
| 5 | 5 | 0.5230E+04 | 6 | 0,1558E+04 | -2 | 0.0 | 4 | 0.9066E+03 | -2 | 0.0 |
| 6 | 6 | 0.5230E+04 | 5 | 0,1558E+04 | 4 | 0.1151E+04 | 6 | 0.7502E+03 | -2 | 0.0 |
| 7 | 4 | 0.7627E+04 | 5 | 0,1558E+04 | -2 | 0.0 | 8 | 0.8919E+03 | 8 | 0.5098E+03 |
| 8 | 3 | 0.6021E+04 | 5 | 0,1558E+04 | 4 | 0.1151E+04 | -2 | 0.0 | -2 | 0.0 |
| 9 | 5 | 0.5230E+04 | 6 | 0,1558E+04 | 4 | 0.1151E+04 | -2 | 0.0 | -2 | 0.0 |
| 10 | 5 | 0.5230E+04 | 5 | 0,1558E+04 | 5 | 0.1151E+04 | 7 | 0.7447E+03 | -2 | 0.0 |
| 11 | 4 | 0.5230E+04 | 7 | 0,1558E+04 | 6 | 0.1151E+04 | 5 | 0.8919E+03 | -2 | 0.0 |
| 12 | 4 | 0.5230E+04 | 6 | 0,1558E+04 | 6 | 0.1314E+04 | -2 | 0.0 | -2 | 0.0 |
| 13 | 4 | 0.5230E+04 | 4 | 0,1558E+04 | 5 | 0.1138E+04 | 4 | 0.9066E+03 | 5 | 0.9232E+03 |
| 14 | 6 | 0.5230E+04 | 4 | 0,1558E+04 | 4 | 0.1311E+04 | -2 | 0.0 | 5 | 0.7605E+03 |
| 15 | 4 | 0.5230E+04 | 5 | 0,1558E+04 | 5 | 0.1151E+04 | 6 | 0.7548E+03 | -2 | 0.0 |
| 16 | 5 | 0.6021E+04 | 4 | 0,1558E+04 | 4 | 0.1314E+04 | -2 | 0.0 | -2 | 0.0 |
| 17 | 5 | 0.5230E+04 | 3 | 0,1558E+04 | -2 | 0.0 | -2 | 0.0 | -2 | 0.0 |
| 18 | 5 | 0.6021E+04 | 5 | 0,1558E+04 | 6 | 0.1123E+04 | -2 | 0.0 | -2 | 0.0 |
| 19 | 4 | 0.5230E+04 | 5 | 0,1558E+04 | 4 | 0.1151E+04 | 6 | 0.7922E+03 | -2 | 0.0 |
| 20 | 6 | 0.6021E+04 | 5 | 0,1558E+04 | 5 | 0.1151E+04 | -2 | 0.0 | 8 | 0.5107E+03 |

*Table 11.7*

| | | 2 | | 3 | | 4 | | 5 | | 6 |
|---|---|---|---|---|---|---|---|---|---|---|
| 1 | 3 | 0,1128E+05 | 4 | 0,6887E+04 | 4 | 0,4303E+04 | 4 | 0,3649E+04 | 3 | 0,2970E+04 |
| 2 | 3 | 0,1111E+05 | 3 | 0,6721E+04 | 5 | 0,4268E+04 | 6 | 0,3374E+04 | 3 | 0,2867E+04 |
| 3 | 6 | 0,1111E+05 | 2 | 0,6630E+04 | 3 | 0,5803E+04 | 5 | 0,3649E+04 | 2 | 0,2967E+04 |
| 4 | 3 | 0,1030E+05 | 3 | 0,6865E+04 | 4 | 0,4294E+04 | 3 | 0,3374E+04 | 3 | 0,2967E+04 |
| 5 | 3 | 0,1030E+05 | 4 | 0,6721E+04 | 4 | 0,4266E+04 | 5 | 0,3374E+04 | 2 | 0,2970E+04 |
| 6 | 3 | 0,1111E+05 | 3 | 0,7090E+04 | 5 | 0,4268E+04 | 2 | 0,3536E+04 | 3 | 0,3050E+04 |
| 7 | 3 | 0,1128E+05 | 3 | 0,6721E+04 | 4 | 0,4266E+04 | 3 | 0,3374E+04 | 5 | 0,3058E+04 |
| 8 | 3 | 0,1111E+05 | 5 | 0,6721E+04 | 4 | 0,4266E+04 | 3 | 0,3374E+04 | 3 | 0,2967E+04 |
| 9 | 3 | 0,1030E+05 | 3 | 0,6887E+04 | 4 | 0,4268E+04 | 5 | 0,3609E+04 | 5 | 0,2921E+04 |
| 10 | 2 | 0,1128E+05 | 4 | 0,6887E+04 | 4 | 0,4266E+04 | 4 | 0,3536E+04 | 3 | 0,2867E+04 |
| 11 | 2 | 0,1109E+05 | 7 | 0,6721E+04 | 3 | 0,4266E+04 | 4 | 0,3374E+04 | 3 | 0,2970E+04 |
| 12 | 2 | 0,1109E+05 | 3 | 0,7090E+04 | 4 | 0,4294E+04 | 4 | 0,3730E+04 | 3 | 0,2970E+04 |
| 13 | 2 | 0,1030E+05 | 4 | 0,6865E+04 | 4 | 0,4268E+04 | 5 | 0,3596E+04 | 3 | 0,2970E+04 |
| 14 | 5 | 0,1111E+05 | 3 | 0,6887E+04 | 2 | 0,5758E+04 | 3 | 0,3374E+04 | 3 | 0,2970E+04 |
| 15 | 3 | 0,1111E+05 | 4 | 0,7090E+04 | 5 | 0,4294E+04 | 4 | 0,3374E+04 | 4 | 0,2967E+04 |
| 16 | 2 | 0,1030E+05 | 2 | 0,6630E+04 | 4 | 0,4294E+04 | 3 | 0,3374E+04 | 3 | 0,2967E+04 |
| 17 | 2 | 0,1030E+05 | 6 | 0,6721E+04 | 4 | 0,4268E+04 | 3 | 0,3649E+04 | 3 | 0,2979E+04 |
| 18 | 3 | 0,1030E+05 | 5 | 0,6721E+04 | 3 | 0,4303E+04 | 4 | 0,3649E+04 | 4 | 0,2867E+04 |
| 19 | 2 | 0,1109E+05 | 6 | 0,6721E+04 | 3 | 0,4266E+04 | 3 | 0,3536E+04 | 2 | 0,2970E+04 |
| 20 | 5 | 0,1111E+05 | 2 | 0,6630E+04 | 4 | 0,4294E+04 | 3 | 0,3649E+04 | 2 | 0,2978E+04 |

| | | 2 | | 3 | | 4 | | 5 | | 6 |
|---|---|---|---|---|---|---|---|---|---|---|
| 1 | 5 | 0.1034E+05 | 4 | 0.6721E+04 | 9 | 0.4323E+04 | 6 | 0.3374E+04 | 5 | 0.3283E+04 |
| 2 | 3 | 0.1146E+05 | 5 | 0.7190E+04 | 12 | 0.4405E+04 | 7 | 0.3374E+04 | 7 | 0.3012E+04 |
| 3 | 3 | 0.1146E+05 | 4 | 0.6630E+04 | 7 | 0.4405E+04 | 6 | 0.3435E+04 | -2 | 0.0 |
| 4 | 4 | 0.1031E+05 | 5 | 0.6925E+04 | 9 | 0.4405E+04 | 5 | 0.3657E+04 | -2 | 0.0 |
| 5 | 3 | 0.1128E+05 | 5 | 0.6829E+04 | 4 | 0.4305E+04 | 3 | 0.4029E+04 | 6 | 0.3049E+04 |
| 6 | 4 | 0.1111E+05 | 6 | 0.6999E+04 | 7 | 0.4585E+04 | 5 | 0.3821E+04 | 6 | 0.3064E+04 |
| 7 | 3 | 0.1130E+05 | 6 | 0.6682E+04 | 10 | 0.4405E+04 | 7 | 0.3374E+04 | 7 | 0.3141E+04 |
| 8 | 3 | 0.1109E+05 | 4 | 0.6856E+04 | 5 | 0.6317E+04 | 5 | 0.3561E+04 | -2 | 0.0 |
| 9 | 3 | 0.1128E+05 | 4 | 0.7093E+04 | 4 | 0.6448E+04 | 6 | 0.5386E+04 | 6 | 0.2955E+04 |
| 10 | 5 | 0.1111E+05 | 7 | 0.6887E+04 | 8 | 0.4585E+04 | 7 | 0.3780E+04 | 4 | 0.3035E+04 |
| 11 | 3 | 0.1130E+05 | 5 | 0.7090E+04 | 4 | 0.4287E+04 | 6 | 0.3374E+04 | 4 | 0.3336E+04 |
| 12 | 6 | 0.1034E+05 | 4 | 0.6738E+04 | 8 | 0.4591E+04 | 4 | 0.5287E+04 | -2 | 0.0 |
| 13 | 6 | 0.1031E+05 | 4 | 0.6890E+04 | 5 | 0.4585E+04 | 6 | 0.3596E+04 | 7 | 0.3083E+04 |
| 14 | 5 | 0.1111E+05 | 4 | 0.6757E+04 | 6 | 0.4585E+04 | 5 | 0.3690E+04 | 7 | 0.3002E+04 |
| 15 | 3 | 0.1111E+05 | 4 | 0.6886E+04 | 5 | 0.4405E+04 | 6 | 0.3780E+04 | 7 | 0.3179E+04 |
| 16 | 6 | 0.1034E+05 | 6 | 0.6637E+04 | 11 | 0.4405E+04 | 5 | 0.3821E+04 | 5 | 0.3122E+04 |
| 17 | 3 | 0.1034E+05 | 4 | 0.6886E+04 | 10 | 0.4298E+04 | 8 | 0.3374E+04 | -2 | 0.0 |
| 18 | 3 | 0.1130E+05 | 4 | 0.7188E+04 | 4 | 0.4280E+04 | 6 | 0.3374E+04 | 5 | 0.3178E+04 |
| 19 | 3 | 0.1116E+05 | 4 | 0.6829E+04 | 4 | 0.4330E+04 | 4 | 0.3888E+04 | 5 | 0.3017E+04 |
| 20 | 5 | 0.1111E+05 | 5 | 0.6856E+04 | 7 | 0.4306E+04 | 6 | 0.3374E+04 | 4 | 0.3014E+04 |

| | | 2 | | 3 | | 4 | | 5 | | 6 |
|---|---|---|---|---|---|---|---|---|---|---|
| 1 | 2 | 0.1030E+05 | 0 | 0.6721E+04 | 1 | 0.4303E+04 | 0 | 0.3374E+04 | 2 | 0.3050E+04 |
| 2 | 3 | 0.1111E+05 | 5 | 0.6721E+04 | 1 | 0.4294E+04 | 0 | 0.3374E+04 | 2 | 0.2969E+04 |
| 3 | 3 | 0.1111E+05 | 0 | 0.6630E+04 | 1 | 0.4294E+04 | 1 | 0.3374E+04 | -2 | 0.0 |
| 4 | 1 | 0.1030E+05 | 1 | 0.6887E+04 | 1 | 0.4294E+04 | 1 | 0.3652E+04 | -2 | 0.0 |
| 5 | 0 | 0.1128E+05 | 2 | 0.6721E+04 | 1 | 0.4266E+04 | 3 | 0.3639E+04 | 1 | 0.2993E+04 |
| 6 | 0 | 0.1111E+05 | 5 | 0.6721E+04 | 2 | 0.4268E+04 | 1 | 0.3374E+04 | 4 | 0.2954E+04 |
| 7 | 1 | 0.1128E+05 | 2 | 0.6630E+04 | 1 | 0.4294E+04 | 0 | 0.3374E+04 | 1 | 0.2921E+04 |
| 8 | 0 | 0.1109E+05 | 3 | 0.6721E+04 | 3 | 0.4294E+04 | 2 | 0.3374E+04 | -2 | 0.0 |
| 9 | 0 | 0.1128E+05 | 1 | 0.7090E+04 | 3 | 0.4303E+04 | 4 | 0.3780E+04 | 1 | 0.2921E+04 |
| 10 | 0 | 0.1111E+05 | 0 | 0.6887E+04 | 2 | 0.4268E+04 | 2 | 0.3596E+04 | 1 | 0.2978E+04 |
| 11 | 1 | 0.1128E+05 | 0 | 0.7090E+04 | 2 | 0.4268E+04 | 0 | 0.3374E+04 | 3 | 0.3001E+04 |
| 12 | 2 | 0.1030E+05 | 1 | 0.6721E+04 | 2 | 0.4268E+04 | 1 | 0.5212E+04 | -2 | 0.0 |
| 13 | 1 | 0.1030E+05 | 2 | 0.6865E+04 | 2 | 0.4268E+04 | 0 | 0.3596E+04 | 1 | 0.3058E+04 |
| 14 | 0 | 0.1111E+05 | 1 | 0.6721E+04 | 2 | 0.4268E+04 | 1 | 0.3629E+04 | 2 | 0.2887E+04 |
| 15 | 0 | 0.1111E+05 | 1 | 0.6865E+04 | 1 | 0.4294E+04 | 0 | 0.3780E+04 | 2 | 0.2921E+04 |
| 16 | 2 | 0.1030E+05 | 1 | 0.6630E+04 | 1 | 0.4294E+04 | 1 | 0.3374E+04 | 2 | 0.3045E+04 |
| 17 | 2 | 0.1030E+05 | 3 | 0.6630E+04 | 1 | 0.4294E+04 | 0 | 0.3374E+04 | -2 | 0.0 |
| 18 | 1 | 0.1128E+05 | 2 | 0.7090E+04 | 1 | 0.4268E+04 | 0 | 0.3374E+04 | 1 | 0.2921E+04 |
| 19 | 0 | 0.1116E+05 | 2 | 0.6721E+04 | 1 | 0.4303E+04 | 2 | 0.3596E+04 | 1 | 0.2954E+04 |
| 20 | 0 | 0.1111E+05 | 3 | 0.6721E+04 | 1 | 0.4303E+04 | 0 | 0.3374E+04 | 1 | 0.2978E+04 |

*Table 11.8*

| | 2 | 3 | 4 | 5 | 6 |
|---|---|---|---|---|---|
| 1 | 2 0.5596E+04 | 3 0.3322E+04 | 5 0.2514E+04 | 5 0.1963E+04 | 4 0.1750E+04 |
| 2 | 2 0.5596E+04 | 4 0.3322E+04 | 5 0.2514E+04 | 5 0.1980E+04 | 3 0.1750E+04 |
| 3 | 2 0.5596E+04 | 2 0.3322E+04 | 2 0.2532E+04 | 3 0.2043E+04 | 3 0.1651E+04 |
| 4 | 2 0.5596E+04 | 4 0.3322E+04 | 3 0.2532E+04 | 2 0.1963E+04 | 2 0.1686E+04 |
| 5 | 2 0.5596E+04 | 3 0.3322E+04 | 3 0.2532E+04 | 3 0.2038E+04 | 5 0.1691E+04 |
| 6 | 2 0.5596E+04 | 3 0.3322E+04 | 3 0.2532E+04 | 3 0.2043E+04 | 3 0.1750E+04 |
| 7 | 2 0.5596E+04 | 3 0.3322E+04 | 3 0.2532E+04 | 2 0.1993E+04 | 5 0.1640E+04 |
| 8 | 2 0.5596E+04 | 2 0.3322E+04 | 2 0.2532E+04 | 6 0.1963E+04 | 3 0.1640E+04 |
| 9 | 2 0.5596E+04 | 3 0.3322E+04 | 2 0.2624E+04 | 3 0.1963E+04 | 4 0.1617E+04 |
| 10 | 2 0.5596E+04 | 2 0.3322E+04 | 3 0.2514E+04 | 3 0.2096E+04 | 3 0.1641E+04 |
| 11 | 2 0.5596E+04 | 5 0.3322E+04 | 3 0.2532E+04 | 2 0.1963E+04 | 4 0.1750E+04 |
| 12 | 2 0.5596E+04 | 3 0.3322E+04 | 3 0.2514E+04 | 3 0.2043E+04 | 2 0.1609E+04 |
| 13 | 2 0.5596E+04 | 3 0.3322E+04 | 3 0.2514E+04 | 2 0.1993E+04 | 2 0.1622E+04 |
| 14 | 2 0.5596E+04 | 3 0.3322E+04 | 3 0.2514E+04 | 4 0.1963E+04 | 7 0.1691E+04 |
| 15 | 3 0.5596E+04 | 2 0.3322E+04 | 3 0.2532E+04 | 3 0.1993E+04 | 3 0.1651E+04 |
| 16 | 2 0.5596E+04 | 2 0.3322E+04 | 3 0.2532E+04 | 3 0.2038E+04 | 3 0.1651E+04 |
| 17 | 1 0.5596E+04 | 4 0.3322E+04 | 3 0.2514E+04 | 4 0.1980E+04 | 2 0.1651E+04 |
| 18 | 2 0.5596E+04 | 3 0.3322E+04 | 3 0.2514E+04 | 5 0.1980E+04 | 3 0.1609E+04 |
| 19 | 2 0.5596E+04 | 3 0.3322E+04 | 3 0.2532E+04 | 4 0.1963E+04 | 3 0.1609E+04 |
| 20 | 2 0.5596E+04 | 2 0.3322E+04 | 5 0.2514E+04 | 3 0.1993E+04 | 3 0.1609E+04 |

| | 2 | 3 | 4 | 5 | 6 |
|---|---|---|---|---|---|
| 1 | 5 0.5596E+04 | 5 0.3970E+04 | 5 0.3052E+04 | -2 0.0 | 7 0.1842E+04 |
| 2 | 5 0.5596E+04 | 6 0.3970E+04 | 5 0.3597E+04 | 7 0.2225E+04 | -2 0.0 |
| 3 | 6 0.5596E+04 | 5 0.3970E+04 | 7 0.3506E+04 | 7 0.1980E+04 | -2 0.0 |
| 4 | 6 0.5596E+04 | 9 0.3970E+04 | 4 0.2583E+04 | 5 0.2257E+04 | 7 0.1777E+04 |
| 5 | 5 0.5596E+04 | 5 0.3322E+04 | -2 0.0 | 8 0.1984E+04 | 4 0.1970E+04 |
| 6 | 5 0.5596E+04 | 6 0.3322E+04 | 5 0.2532E+04 | 6 0.2138E+04 | 5 0.1922E+04 |
| 7 | 5 0.5596E+04 | 7 0.3970E+04 | 5 0.2536E+04 | 7 0.2501E+04 | 4 0.2135E+04 |
| 8 | 5 0.5596E+04 | 8 0.3970E+04 | -2 0.0 | 5 0.2914E+04 | 6 0.1648E+04 |
| 9 | 5 0.5596E+04 | 13 0.3970E+04 | 8 0.2898E+04 | 5 0.2467E+04 | 6 0.1735E+04 |
| 10 | 5 0.5596E+04 | 6 0.3970E+04 | 9 0.2898E+04 | 8 0.2174E+04 | 6 0.1842E+04 |
| 11 | 5 0.5596E+04 | 5 0.4879E+04 | 5 0.2583E+04 | -2 0.0 | 7 0.1650E+04 |
| 12 | 5 0.5596E+04 | 12 0.3970E+04 | 6 0.3506E+04 | 5 0.2122E+04 | 7 0.1674E+04 |
| 13 | 5 0.5596E+04 | 8 0.3970E+04 | 9 0.2854E+04 | 6 0.2467E+04 | -2 0.0 |
| 14 | 5 0.5596E+04 | 6 0.3970E+04 | 9 0.2898E+04 | 5 0.2500E+04 | 4 0.1750E+04 |
| 15 | 7 0.5596E+04 | 4 0.3970E+04 | -2 0.0 | 16 0.2069E+04 | 6 0.1667E+04 |
| 16 | 5 0.5596E+04 | 5 0.5076E+04 | 8 0.2532E+04 | 8 0.1984E+04 | -2 0.0 |
| 17 | 5 0.5596E+04 | 6 0.3970E+04 | 7 0.2514E+04 | 9 0.2125E+04 | 6 0.1740E+04 |
| 18 | 5 0.5596E+04 | 7 0.3546E+04 | 16 0.2594E+04 | -2 0.0 | 6 0.1674E+04 |
| 19 | 5 0.5596E+04 | 6 0.3322E+04 | 6 0.3018E+04 | 5 0.2368E+04 | -2 0.0 |
| 20 | 5 0.5596E+04 | 9 0.3970E+04 | 9 0.2898E+04 | -2 0.0 | 9 0.1674E+04 |

| | 2 | 3 | 4 | 5 | 6 |
|---|---|---|---|---|---|
| 1 | 0 0.5596E+04 | 2 0.3322E+04 | 3 0.2514E+04 | -2 0.0 | 2 0.1622E+04 |
| 2 | 0 0.5596E+04 | 2 0.3322E+04 | 2 0.2532E+04 | 4 0.2096E+04 | -2 0.0 |
| 3 | 0 0.5596E+04 | 2 0.3322E+04 | 3 0.2514E+04 | 0 0.1980E+04 | -2 0.0 |
| 4 | 0 0.5596E+04 | 2 0.3322E+04 | 1 0.2514E+04 | 4 0.2096E+04 | 2 0.1640E+04 |
| 5 | 0 0.5596E+04 | 0 0.3322E+04 | -2 0.0 | 0 0.1984E+04 | 4 0.1622E+04 |
| 6 | 0 0.5596E+04 | 0 0.3322E+04 | 0 0.2532E+04 | 2 0.2065E+04 | 3 0.1622E+04 |
| 7 | 0 0.5596E+04 | 2 0.3322E+04 | 2 0.2514E+04 | 3 0.2096E+04 | 6 0.1788E+04 |
| 8 | 0 0.5596E+04 | 2 0.3322E+04 | -2 0.0 | 5 0.2038E+04 | 2 0.1617E+04 |
| 9 | 0 0.5596E+04 | 2 0.3322E+04 | 3 0.2514E+04 | 3 0.2096E+04 | 5 0.1609E+04 |
| 10 | 0 0.5596E+04 | 2 0.3322E+04 | 3 0.2514E+04 | 2 0.2096E+04 | 5 0.1617E+04 |
| 11 | 0 0.5596E+04 | 3 0.3322E+04 | 1 0.2514E+04 | -2 0.0 | 0 0.1650E+04 |
| 12 | 0 0.5596E+04 | 2 0.3322E+04 | 4 0.2514E+04 | 0 0.2122E+04 | 3 0.1622E+04 |
| 13 | 0 0.5596E+04 | 2 0.3322E+04 | 2 0.2514E+04 | 3 0.2096E+04 | -2 0.0 |
| 14 | 0 0.5596E+04 | 2 0.3322E+04 | 3 0.2514E+04 | 5 0.2096E+04 | 8 0.1609E+04 |
| 15 | 0 0.5596E+04 | 2 0.3322E+04 | -2 0.0 | 1 0.2065E+04 | 3 0.1609E+04 |
| 16 | 0 0.5596E+04 | 5 0.3322E+04 | 0 0.2532E+04 | 0 0.1984E+04 | -2 0.0 |
| 17 | 0 0.5596E+04 | 2 0.3322E+04 | 0 0.2514E+04 | 1 0.2122E+04 | 1 0.1689E+04 |
| 18 | 0 0.5596E+04 | 1 0.3322E+04 | 2 0.2514E+04 | -2 0.0 | 3 0.1617E+04 |
| 19 | 0 0.5596E+04 | 0 0.3322E+04 | 6 0.2514E+04 | 2 0.2065E+04 | -2 0.0 |
| 20 | 0 0.5596E+04 | 2 0.3322E+04 | 3 0.2514E+04 | -2 0.0 | 3 0.1617E+04 |

*Table 11.9*

| | | 2 | | 3 | | 4 | | 5 | | 6 |
|---|---|---|---|---|---|---|---|---|---|---|
| 1 | 2 | 0.1709E+05 | 3 | 0.8627E+04 | 4 | 0.5424E+04 | 5 | 0.4810E+04 | 8 | 0.3220E+04 |
| 2 | 2 | 0.1712E+05 | 4 | 0.8627E+04 | 4 | 0.5424E+04 | 4 | 0.4098E+04 | 8 | 0.3220E+04 |
| 3 | 2 | 0.1712E+05 | 3 | 0.8627E+04 | 6 | 0.5424E+04 | 6 | 0.4064E+04 | 4 | 0.3592E+04 |
| 4 | 2 | 0.1709E+05 | 6 | 0.8627E+04 | 6 | 0.5424E+04 | 4 | 0.4801E+04 | 6 | 0.3578E+04 |
| 5 | 2 | 0.1586E+05 | 6 | 0.1019E+05 | 6 | 0.5450E+04 | 2 | 0.4989E+04 | 4 | 0.3737E+04 |
| 6 | 2 | 0.1712E+05 | 7 | 0.8627E+04 | 5 | 0.5450E+04 | 9 | 0.4064E+04 | 5 | 0.3578E+04 |
| 7 | 3 | 0.1709E+05 | 5 | 0.8627E+04 | 5 | 0.8007E+04 | 4 | 0.4794E+04 | 4 | 0.3592E+04 |
| 8 | 2 | 0.1709E+05 | 5 | 0.8627E+04 | 4 | 0.8007E+04 | 6 | 0.4064E+04 | 6 | 0.3220E+04 |
| 9 | 2 | 0.1712E+05 | 7 | 0.8627E+04 | 6 | 0.5424E+04 | 5 | 0.4810E+04 | 3 | 0.3592E+04 |
| 10 | 4 | 0.1712E+05 | 2 | 0.8627E+04 | 3 | 0.5437E+04 | 7 | 0.4935E+04 | 6 | 0.3522E+04 |
| 11 | 4 | 0.1586E+05 | 2 | 0.8627E+04 | 6 | 0.5437E+04 | 5 | 0.4810E+04 | 6 | 0.3220E+04 |
| 12 | 2 | 0.1649E+05 | 3 | 0.8627E+04 | 5 | 0.5437E+04 | 4 | 0.4064E+04 | 4 | 0.3522E+04 |
| 13 | 3 | 0.1712E+05 | 3 | 0.8627E+04 | 5 | 0.5450E+04 | 2 | 0.4810E+04 | 4 | 0.3522E+04 |
| 14 | 2 | 0.1712E+05 | 6 | 0.1019E+05 | 7 | 0.5450E+04 | 6 | 0.3842E+04 | 7 | 0.3522E+04 |
| 15 | 2 | 0.1586E+05 | 6 | 0.1019E+05 | 6 | 0.5424E+04 | 7 | 0.4935E+04 | 6 | 0.3578E+04 |
| 16 | 2 | 0.1712E+05 | 7 | 0.8627E+04 | 6 | 0.5437E+04 | 5 | 0.4810E+04 | 5 | 0.3578E+04 |
| 17 | 2 | 0.1712E+05 | 3 | 0.8627E+04 | 6 | 0.5437E+04 | 3 | 0.4798E+04 | 3 | 0.3592E+04 |
| 18 | 2 | 0.1709E+05 | 5 | 0.8627E+04 | 6 | 0.5450E+04 | 6 | 0.4794E+04 | 5 | 0.3578E+04 |
| 19 | 2 | 0.1649E+05 | 7 | 0.8627E+04 | 5 | 0.5437E+04 | 4 | 0.4798E+04 | 9 | 0.3578E+04 |
| 20 | 3 | 0.1712E+05 | 5 | 0.8627E+04 | 7 | 0.5437E+04 | 5 | 0.4801E+04 | 3 | 0.3592E+04 |

| | | 2 | | 3 | | 4 | | 5 | | 6 |
|---|---|---|---|---|---|---|---|---|---|---|
| 1 | 5 | 0.1900E+05 | 10 | 0.9705E+04 | 6 | 0.8099E+04 | -2 | 0.0 | -2 | 0.0 |
| 2 | 3 | 0.1649E+05 | 8 | 0.1024E+05 | -2 | 0.0 | 6 | 0.5206E+04 | -2 | 0.0 |
| 3 | 3 | 0.1586E+05 | 8 | 0.1024E+05 | -2 | 0.0 | -2 | 0.0 | 7 | 0.2540E+04 |
| 4 | 13 | 0.1709E+05 | 6 | 0.8627E+04 | 4 | 0.5572E+04 | 9 | 0.5080E+04 | -2 | 0.0 |
| 5 | 3 | 0.1586E+05 | 8 | 0.1024E+05 | 5 | 0.5492E+04 | 6 | 0.3870E+04 | 8 | 0.3592E+04 |
| 6 | 4 | 0.1709E+05 | 6 | 0.1019E+05 | 8 | 0.5491E+04 | -2 | 0.0 | -2 | 0.0 |
| 7 | 3 | 0.1649E+05 | 8 | 0.1024E+05 | 8 | 0.5479E+04 | 6 | 0.4819E+04 | -2 | 0.0 |
| 8 | 3 | 0.1648E+05 | 5 | 0.8627E+04 | 6 | 0.5455E+04 | 11 | 0.3937E+04 | -2 | 0.0 |
| 9 | 11 | 0.1709E+05 | 6 | 0.1019E+05 | 11 | 0.5455E+04 | 5 | 0.4845E+04 | 7 | 0.2519E+04 |
| 10 | 4 | 0.1586E+05 | 7 | 0.9705E+04 | 7 | 0.5479E+04 | 8 | 0.3910E+04 | -2 | 0.0 |
| 11 | 5 | 0.1586E+05 | 4 | 0.8650E+04 | 4 | 0.5492E+04 | 4 | 0.3896E+04 | -2 | 0.0 |
| 12 | 4 | 0.1649E+05 | 10 | 0.9705E+04 | -2 | 0.0 | 8 | 0.3873E+04 | -2 | 0.0 |
| 13 | 4 | 0.1709E+05 | 7 | 0.9705E+04 | 6 | 0.5437E+04 | 7 | 0.4103E+04 | -2 | 0.0 |
| 14 | 4 | 0.1900E+05 | 9 | 0.9705E+04 | 9 | 0.5492E+04 | 4 | 0.5137E+04 | -2 | 0.0 |
| 15 | 3 | 0.1648E+05 | 7 | 0.9705E+04 | 6 | 0.5445E+04 | 8 | 0.4874E+04 | 10 | 0.3663E+04 |
| 16 | 3 | 0.1649E+05 | 5 | 0.1024E+05 | 4 | 0.5480E+04 | -2 | 0.0 | -2 | 0.0 |
| 17 | 11 | 0.1709E+05 | 14 | 0.9705E+04 | 6 | 0.5437E+04 | -2 | 0.0 | 7 | 0.3601E+04 |
| 18 | 10 | 0.1709E+05 | 5 | 0.8627E+04 | 5 | 0.5437E+04 | 8 | 0.3896E+04 | -2 | 0.0 |
| 19 | 3 | 0.1649E+05 | 4 | 0.1019E+05 | 6 | 0.5477E+04 | -2 | 0.0 | 6 | 0.2540E+04 |
| 20 | 4 | 0.1712E+05 | 4 | 0.8627E+04 | 10 | 0.5492E+04 | -2 | 0.0 | -2 | 0.0 |

| | | 2 | | 3 | | 4 | | 5 | | 6 |
|---|---|---|---|---|---|---|---|---|---|---|
| 1 | 2 | 0.1899E+05 | 2 | 0.8627E+04 | 3 | 0.5450E+04 | -2 | 0.0 | -2 | 0.0 |
| 2 | 0 | 0.1649E+05 | 2 | 0.1019E+05 | -2 | 0.0 | 2 | 0.4810E+04 | -2 | 0.0 |
| 3 | 0 | 0.1586E+05 | 2 | 0.1019E+05 | -2 | 0.0 | -2 | 0.0 | 1 | 0.2519E+04 |
| 4 | 0 | 0.1709E+05 | 0 | 0.8627E+04 | 2 | 0.5450E+04 | 2 | 0.4794E+04 | -2 | 0.0 |
| 5 | 0 | 0.1586E+05 | 2 | 0.1019E+05 | 1 | 0.5450E+04 | 2 | 0.3842E+04 | 1 | 0.3578E+04 |
| 6 | 0 | 0.1709E+05 | 0 | 0.1019E+05 | 2 | 0.5450E+04 | -2 | 0.0 | -2 | 0.0 |
| 7 | 0 | 0.1649E+05 | 2 | 0.1019E+05 | 2 | 0.5424E+04 | 2 | 0.4810E+04 | -2 | 0.0 |
| 8 | 2 | 0.1586E+05 | 0 | 0.8627E+04 | 2 | 0.5437E+04 | 1 | 0.3936E+04 | -2 | 0.0 |
| 9 | 0 | 0.1709E+05 | 0 | 0.1019E+05 | 2 | 0.5437E+04 | 3 | 0.3842E+04 | 0 | 0.2519E+04 |
| 10 | 0 | 0.1586E+05 | 2 | 0.8627E+04 | 2 | 0.5424E+04 | 3 | 0.3842E+04 | -2 | 0.0 |
| 11 | 0 | 0.1586E+05 | 1 | 0.8627E+04 | 1 | 0.5450E+04 | 1 | 0.3842E+04 | -2 | 0.0 |
| 12 | 0 | 0.1649E+05 | 2 | 0.8627E+04 | -2 | 0.0 | 2 | 0.3842E+04 | -2 | 0.0 |
| 13 | 0 | 0.1709E+05 | 2 | 0.8627E+04 | 0 | 0.5437E+04 | 1 | 0.4098E+04 | -2 | 0.0 |
| 14 | 1 | 0.1899E+05 | 2 | 0.8627E+04 | 1 | 0.5450E+04 | 2 | 0.4794E+04 | -2 | 0.0 |
| 15 | 2 | 0.1586E+05 | 2 | 0.8627E+04 | 2 | 0.5437E+04 | 2 | 0.4794E+04 | 2 | 0.3522E+04 |
| 16 | 0 | 0.1649E+05 | 2 | 0.1019E+05 | 2 | 0.5424E+04 | -2 | 0.0 | -2 | 0.0 |
| 17 | 0 | 0.1709E+05 | 2 | 0.8627E+04 | 0 | 0.5437E+04 | -2 | 0.0 | 2 | 0.3578E+04 |
| 18 | 0 | 0.1709E+05 | 0 | 0.8627E+04 | 0 | 0.5437E+04 | 1 | 0.3842E+04 | -2 | 0.0 |
| 19 | 0 | 0.1649E+05 | 0 | 0.1019E+05 | 3 | 0.5437E+04 | -2 | 0.0 | 1 | 0.2519E+04 |
| 20 | 0 | 0.1712E+05 | 0 | 0.8627E+04 | 1 | 0.5450E+04 | -2 | 0.0 | -2 | 0.0 |

A comparison of the results of the two methods, which have the same storage requirements, shows that the exchange method is preferable. Only in isolated cases (example 4, n = 6) does the minimal distance method produce a smaller value of the objective function when the number of classes is reasonably large. In terms of computation time, the exchange method is at least no slower than the other. Over a total of twelve examples which included the four presented here, the computation times with TRWEXM and TRWMDM were 395 and 425 seconds CPU time on the TR 440. These times refer to the main programs as described, with KPRINT = 1, NRMAX = 20, N1 = 2, N2 = 6. TRWEXM appears even more superior if one bears in mind that TRWMDM frequently broke off unsuccessfully with IFLAG = 2, that is KIT(N, NR) = −2. Hence, if one is only able to try out one or a few starting partitions, then the exchange method is more efficient in the sense that it is more likely to produce a better end partition, using less computer time.

Examining the combination of both methods (bottom section), we can see that the subsequent use of the exchange method in those cases where no empty classes were formed usually led to a considerable improvement over the values of the objective function achieved with the minimal distance method (see centre section). The number of additional passes of the exchange method is low, as shown in the tables, but rises a little with the number of classes. For the twelve examples mentioned in the previous paragraph, the computer time for the combined methods was 502 seconds. With the larger numbers of classes, the combined method did occasionally produce values of the objective function which were lower than those achieved with the exchange method alone.

Overall, we prefer the exchange method without reservation. If several random starting partitions are used, the distribution of the values of the objective function may give clues about the appropriate number of classes, whenever each starting partition leads to almost the same value of the objective function. We therefore regard it as important to use a relatively simple method such as the exchange method, and to apply it to several starting partitions. For this reason, heuristic methods for finding a single 'good' starting partition are hardly worthwhile, nor are much more elaborate methods which have been mentioned previously. One might occasionally get a better end partition (although there are no such theorems based on complexity theory), but one would have to use much more complicated programs which would need more computer time and hence would not permit the same number of starting partitions to be used as would be possible in the same amount of time with the exchange method. One would not end up with as large a sample of end partitions.

## 11.5 Results from the test examples using the variance criterion

For the four sets of data of Table 11.3 and Fig. 11.1, Figs. 11.2 to 11.5 show the best partitions into n = 2, 3, 4, 5 classes, that is, those for which the variance criterion was the lowest among the 20 starting partitions and three methods whose results are listed in Tables 11.6 to 11.9. Most often, the best partitions were produced by the exchange method, which explains why most of the Figs. 11.2 to 11.5 were produced by the program listed in Table 11.5. One exception is the case of example 1 with n = 5 classes.

With example 1 in Fig. 11.2, the variance criterion apparently performs in the same way as if one had done the partitioning intuitively oneself. The example in Fig. 11.6, taken from [A2], shows that this is not always the case. There, the optimal partitions for n = 2, shown by lines drawn around the classes, depend on the distance between the two sets of points [B16].

As expected, a sensible dissection into classes is produced by example 2 (Fig. 11.3). Of course, the convex hulls of the cluster members are disjoint in this as in the other cases.

It is interesting that, for example, for n = 5 in example 2 one cluster is surrounded by the other four, while in example 4, all five clusters are arranged in a star shape. At present, we do not wish to take the results, which also reflect the theoretical properties of the variance criterion, any further, but in the comparison with other criteria we shall return to the figures. The reader should examine the results closely and should also use his own data sets to find out what happens.

## 11.6 Examples for the Determinant Criterion and for the Criterion of Adaptive Distances

The same few sample data sets shown in Fig. 11.1 will also be used for the determinant criterion and for the criterion of adaptive distances, using the corresponding subroutines DETEXM and DWBEXM. To the data we add four further examples which are chosen to illustrate specific properties of the criteria in question but do not claim to be close to reality as are examples 2 and 4. The data for the new examples are listed in Table 11.10, and the corresponding plots produced by PRINT are shown in Fig. 11.7.

Example 5 consists of two or three sets of points, each of which has the shape of ellipsoid (or, as here, an ellipse) with the same axis orientation. These ought to be detected by DETEXM and even more by DWBEXM. TRWEXM cannot do it. Example 6 has three sets of points consisting of ellipsoids with their axes in different directions. TRWEXM and DETEXM have no chance of discovering these. However, DWBEXM must, in theory, be able to do it. Examples 7 and 8 contain two or three ellipsoids which overlap in pairs and have different axis orientations. They are included to illustrate the capabilities of DWBEXM. Of course we shall also show selected results of applying TRWEXM to examples 5 to 8, and of applying DETEXM to examples 6 to 8. However, it

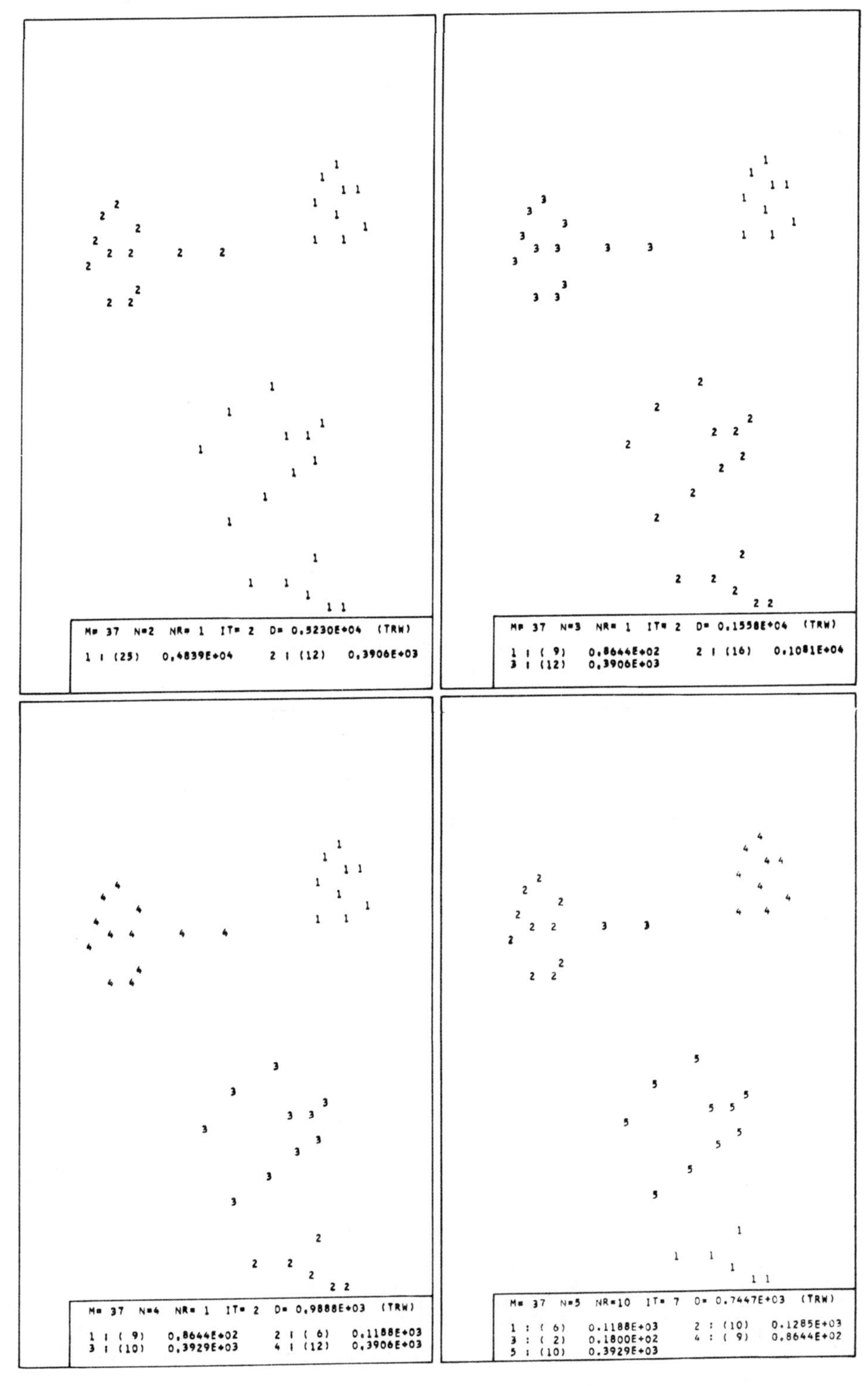
M= 37  N=2  NR= 1  IT= 2  D= 0,5230E+04  (TRW)
1 : (25)  0,4839E+04  2 : (12)  0,3906E+03
M= 37  N=3  NR= 1  IT= 2  D= 0,1558E+04  (TRW)
1 : ( 9)  0,8644E+02  2 : (16)  0,1081E+04
3 : (12)  0,3906E+03
M= 37  N=4  NR= 1  IT= 2  D= 0,9888E+03  (TRW)
1 : ( 9)  0,8644E+02  2 : ( 6)  0,1188E+03
3 : (10)  0,3929E+03  4 : (12)  0,3906E+03
M= 37  N=5  NR=10  IT= 7  D= 0.7447E+03  (TRW)
1 : ( 6)  0.1188E+03  2 : (10)  0.1285E+03
3 : ( 2)  0.1800E+02  4 : ( 9)  0.8644E+02
5 : (10)  0.3929E+03

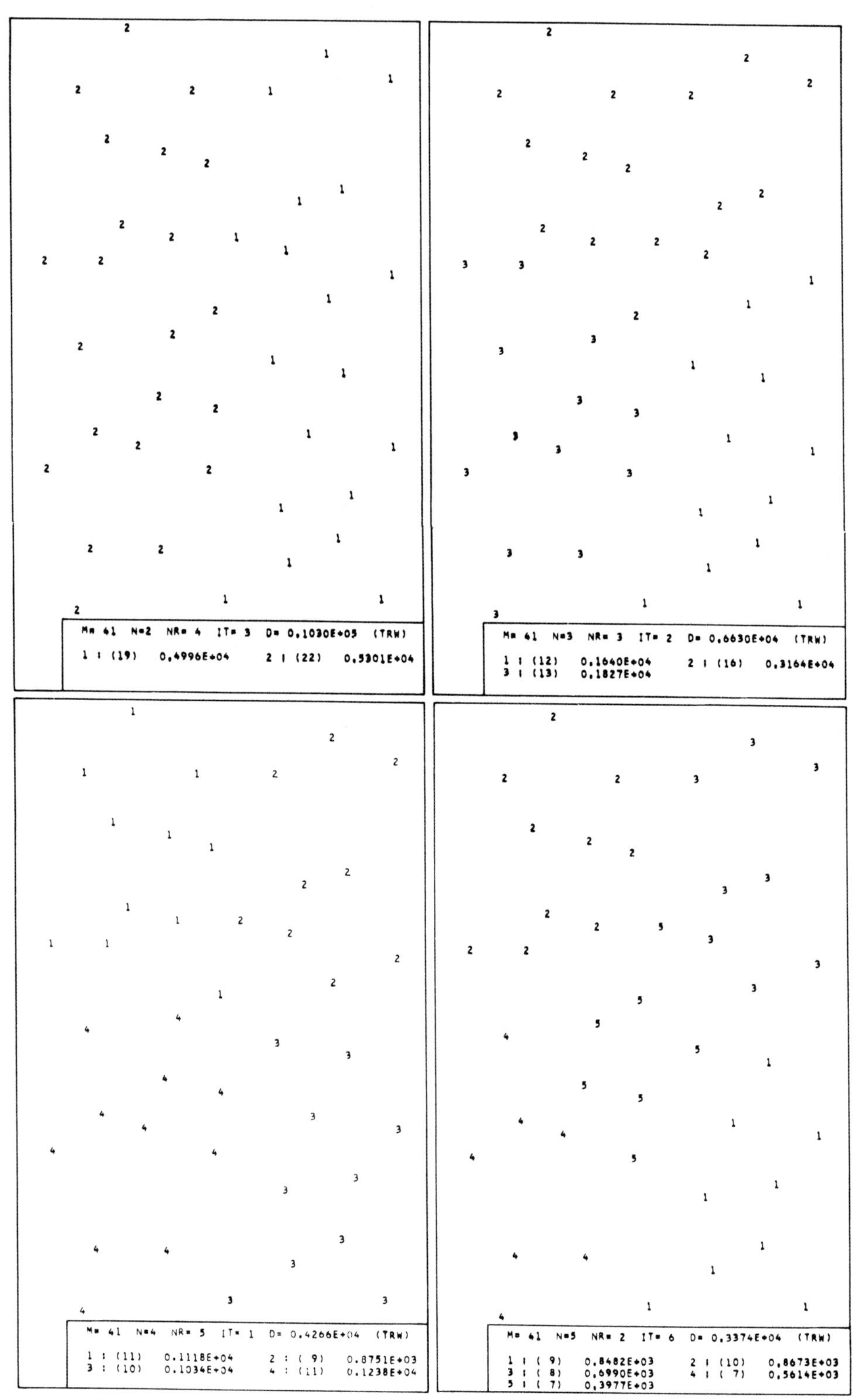
M= 41 N=2 NR= 4 IT= 3 D= 0.1030E+05 (TRW)
1 : (19) 0.4996E+04 2 : (22) 0.5301E+04
M= 41 N=3 NR= 3 IT= 2 D= 0.6630E+04 (TRW)
1 : (12) 0.1640E+04 2 : (16) 0.3164E+04
3 : (13) 0.1827E+04
M= 41 N=4 NR= 5 IT= 1 D= 0.4266E+04 (TRW)
1 : (11) 0.1118E+04 2 : ( 9) 0.8751E+03
3 : (10) 0.1034E+04 4 : (11) 0.1238E+04
M= 41 N=5 NR= 2 IT= 6 D= 0.3374E+04 (TRW)
1 : ( 9) 0.8482E+03 2 : (10) 0.8673E+03
3 : ( 8) 0.6990E+03 4 : ( 7) 0.5614E+03
5 : ( 7) 0.3977E+03

*Fig. 11.3*

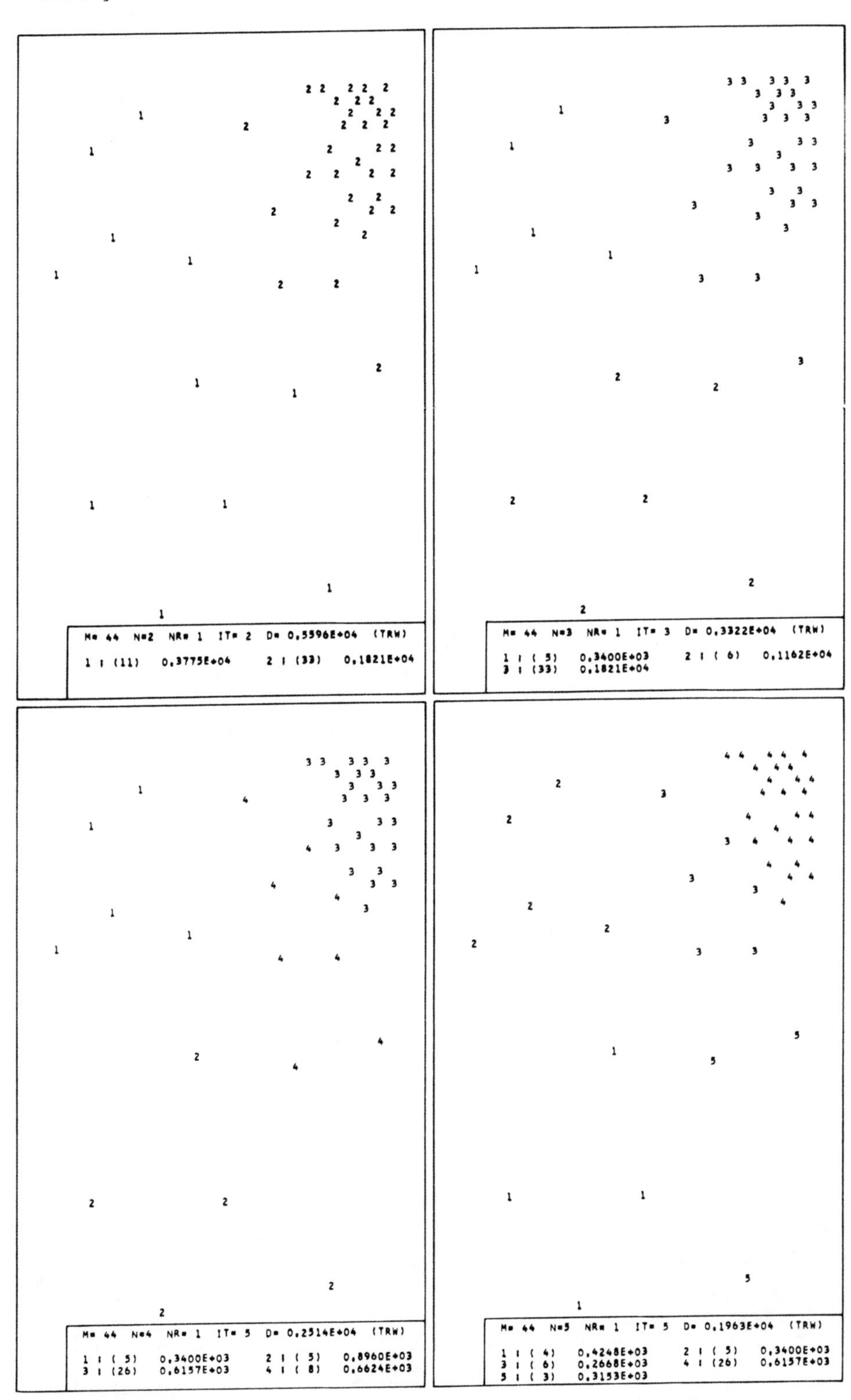
M= 44 N=2 NR= 1 IT= 2 D= 0,5596E+04 (TRW)
1 I (11) 0,3775E+04 2 I (33) 0,1821E+04
M= 44 N=3 NR= 1 IT= 3 D= 0,3322E+04 (TRW)
1 I ( 5) 0,3400E+03 2 I ( 6) 0,1162E+04
3 I (33) 0,1821E+04
M= 44 N=4 NR= 1 IT= 5 D= 0,2514E+04 (TRW)
1 I ( 5) 0,3400E+03 2 I ( 5) 0,8960E+03
3 I (26) 0,6157E+03 4 I ( 8) 0,6624E+03
M= 44 N=5 NR= 1 IT= 5 D= 0,1963E+04 (TRW)
1 I ( 4) 0,4248E+03 2 I ( 5) 0,3400E+03
3 I ( 6) 0,2668E+03 4 I (26) 0,6157E+03
5 I ( 3) 0,3153E+03

*Fig. 11.4*

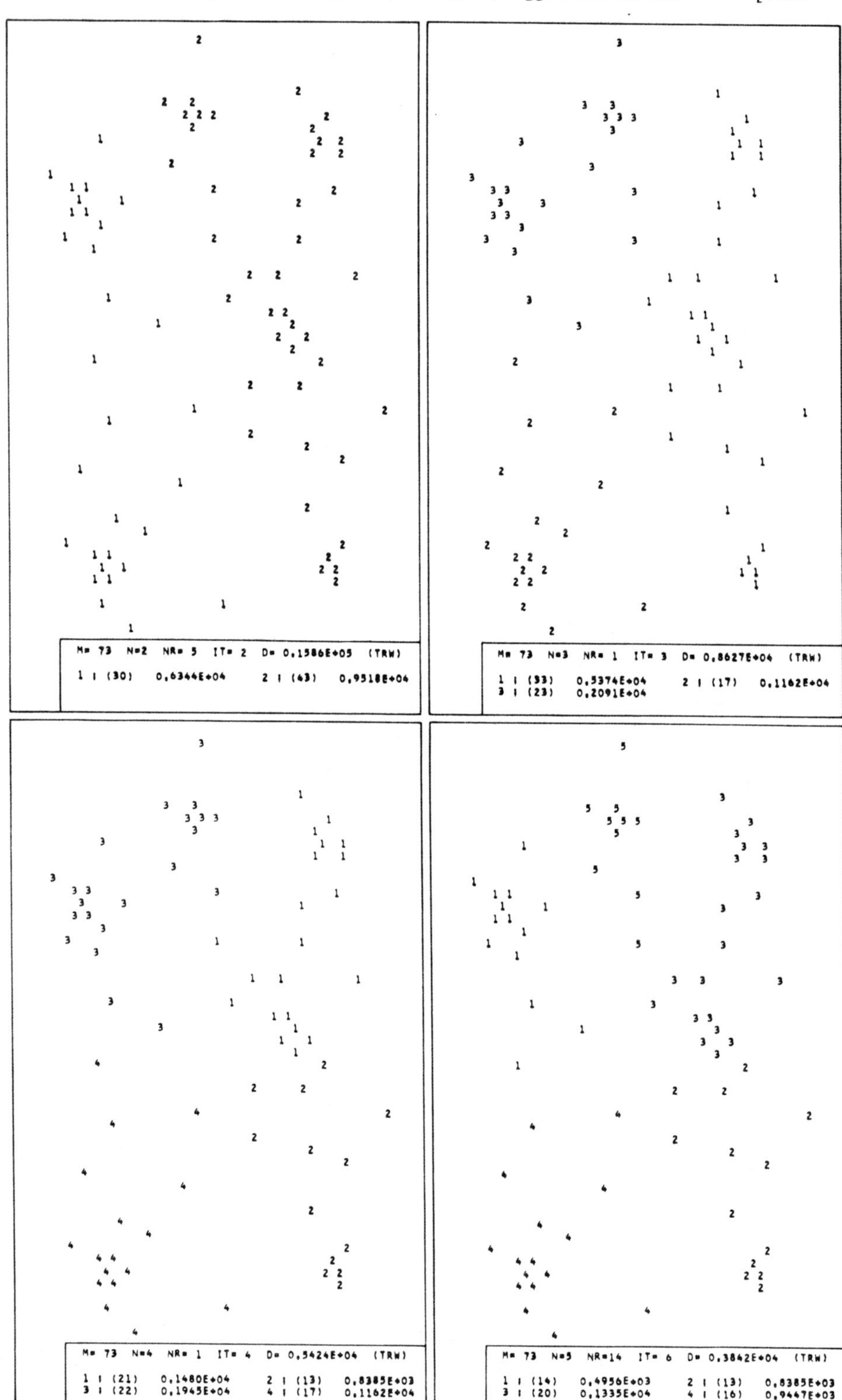
M= 73 N=2 NR= 5 IT= 2 D= 0,1586E+05 (TRW)
1 I (30) 0,6344E+04 2 I (43) 0,9518E+04
M= 73 N=3 NR= 1 IT= 3 D= 0,8627E+04 (TRW)
1 I (33) 0,5374E+04 2 I (17) 0,1162E+04
3 I (23) 0,2091E+04
M= 73 N=4 NR= 1 IT= 4 D= 0,5424E+04 (TRW)
1 I (21) 0,1480E+04 2 I (13) 0,8385E+03
3 I (22) 0,1945E+04 4 I (17) 0,1162E+04
M= 73 N=5 NR=14 IT= 6 D= 0,3842E+04 (TRW)
1 I (14) 0,4956E+03 2 I (13) 0,8385E+03
3 I (20) 0,1335E+04 4 I (16) 0,9447E+03
5 I (10) 0,2282E+03

*Fig. 11.5*

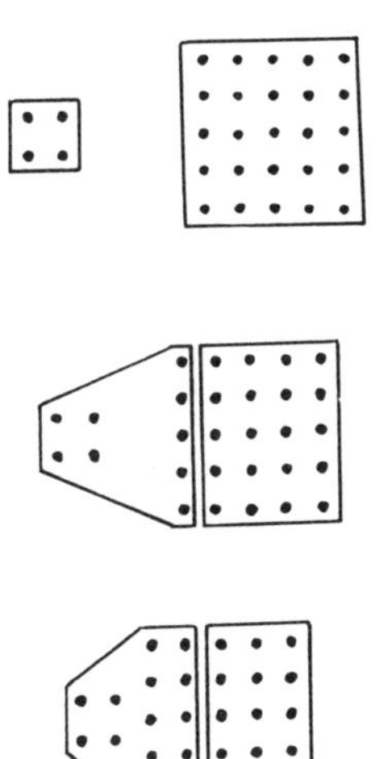

*Fig. 11.6*

| | | | | | | | | | | | | | | | |
|---|---|---|---|---|---|---|---|---|---|---|---|---|---|---|---|
| 45 | 22 | 20 | 18 | 21 | 46 | 19 | 41 | 15 | 17 | 19 | 24 | 14 | 17 | 23 | 42 |
| 21 | 38 | 14 | 12 | 21 | 19 | 34 | 13 | 11 | 19 | 39 | 17 | 35 | 9 | 15 | 17 |
| 9 | 32 | 15 | 29 | 13 | 35 | 11 | 13 | 32 | 24 | 11 | 27 | 9 | 7 | 20 | 28 |
| 6 | 23 | 18 | 22 | 17 | 12 | 9 | | | | | | | | | |
| | | | | | | | | | | | | | | | |
| 50 | 49 | 48 | 46 | 46 | 46 | 44 | 44 | 42 | 42 | 42 | 41 | 39 | 39 | 39 | 39 |
| 38 | 38 | 37 | 36 | 36 | 35 | 35 | 34 | 33 | 33 | 33 | 32 | 32 | 31 | 30 | 30 |
| 29 | 29 | 28 | 28 | 27 | 27 | 25 | 25 | 25 | 24 | 23 | 22 | 21 | 18 | 17 | 17 |
| 16 | 16 | 12 | 11 | 7 | 6 | 1 | | | | | | | | | |
| | | | | | | | | | | | | | | | |
| 25 | 23 | 25 | 21 | 22 | 18 | 46 | 16 | 19 | 43 | 41 | 14 | 16 | 38 | 34 | 37 |
| 16 | 36 | 12 | 14 | 31 | 11 | 28 | 30 | 12 | 9 | 22 | 25 | 8 | 19 | 22 | 19 |
| 16 | 14 | 11 | 6 | 9 | 12 | 17 | 9 | 21 | 25 | 15 | 29 | 19 | 24 | 28 | 32 |
| 37 | 41 | | | | | | | | | | | | | | |
| | | | | | | | | | | | | | | | |
| 42 | 40 | 40 | 39 | 37 | 36 | 35 | 34 | 34 | 34 | 32 | 31 | 31 | 31 | 29 | 29 |
| 28 | 28 | 27 | 27 | 27 | 25 | 25 | 24 | 23 | 22 | 22 | 22 | 20 | 20 | 20 | 18 |
| 17 | 16 | 15 | 13 | 12 | 8 | 8 | 7 | 7 | 7 | 6 | 6 | 5 | 5 | 4 | 4 |
| 3 | 3 | | | | | | | | | | | | | | |
| | | | | | | | | | | | | | | | |
| 11 | 16 | 23 | 13 | 18 | 26 | 17 | 27 | 34 | 23 | 31 | 22 | 37 | 42 | 27 | 34 |
| 37 | 40 | 24 | 39 | 32 | 35 | 37 | 26 | 29 | 43 | 33 | 23 | 27 | 38 | 19 | 29 |
| 13 | 25 | 34 | 19 | 45 | 15 | 22 | 7 | 10 | 18 | 38 | 45 | 15 | 44 | 3 | 5 |
| 11 | 49 | 1 | 5 | | | | | | | | | | | | |
| | | | | | | | | | | | | | | | |
| 50 | 48 | 47 | 46 | 43 | 43 | 39 | 39 | 39 | 38 | 35 | 34 | 34 | 34 | 33 | 32 |
| 31 | 31 | 29 | 29 | 28 | 28 | 27 | 26 | 25 | 25 | 24 | 23 | 23 | 23 | 22 | 22 |
| 21 | 20 | 20 | 19 | 19 | 18 | 18 | 17 | 16 | 16 | 16 | 16 | 13 | 13 | 12 | 12 |
| 12 | 11 | 9 | 8 | | | | | | | | | | | | |
| | | | | | | | | | | | | | | | |
| 16 | 37 | 40 | 20 | 37 | 34 | 38 | 21 | 35 | 24 | 32 | 32 | 24 | 30 | 27 | 30 |
| 25 | 28 | 27 | 1 | 29 | 25 | 3 | 23 | 29 | 24 | 2 | 6 | 19 | 22 | 32 | 10 |
| 20 | 5 | 12 | 16 | 34 | 9 | 15 | 13 | 18 | 15 | 13 | 22 | 11 | 15 | 18 | 24 |
| 36 | 39 | 12 | 10 | 15 | 19 | 24 | 29 | 37 | 9 | 7 | 22 | 26 | 41 | 29 | 34 |
| 39 | 6 | 4 | 26 | 37 | 42 | 44 | 32 | 2 | 37 | 47 | 1 | 30 | 41 | 44 | 46 |
| | | | | | | | | | | | | | | | |
| 50 | 50 | 50 | 49 | 48 | 46 | 45 | 44 | 43 | 42 | 42 | 39 | 38 | 38 | 36 | 35 |
| 34 | 34 | 32 | 31 | 31 | 30 | 29 | 29 | 29 | 28 | 26 | 26 | 26 | 26 | 26 | 25 |
| 24 | 22 | 22 | 22 | 22 | 20 | 20 | 19 | 19 | 18 | 17 | 17 | 15 | 15 | 15 | 15 |
| 15 | 14 | 13 | 12 | 12 | 12 | 12 | 12 | 11 | 10 | 9 | 9 | 9 | 9 | 8 | 8 |
| 8 | 7 | 5 | 5 | 5 | 5 | 5 | 4 | 3 | 2 | 2 | 1 | 1 | 1 | 1 | 1 |

*Table 11.10*

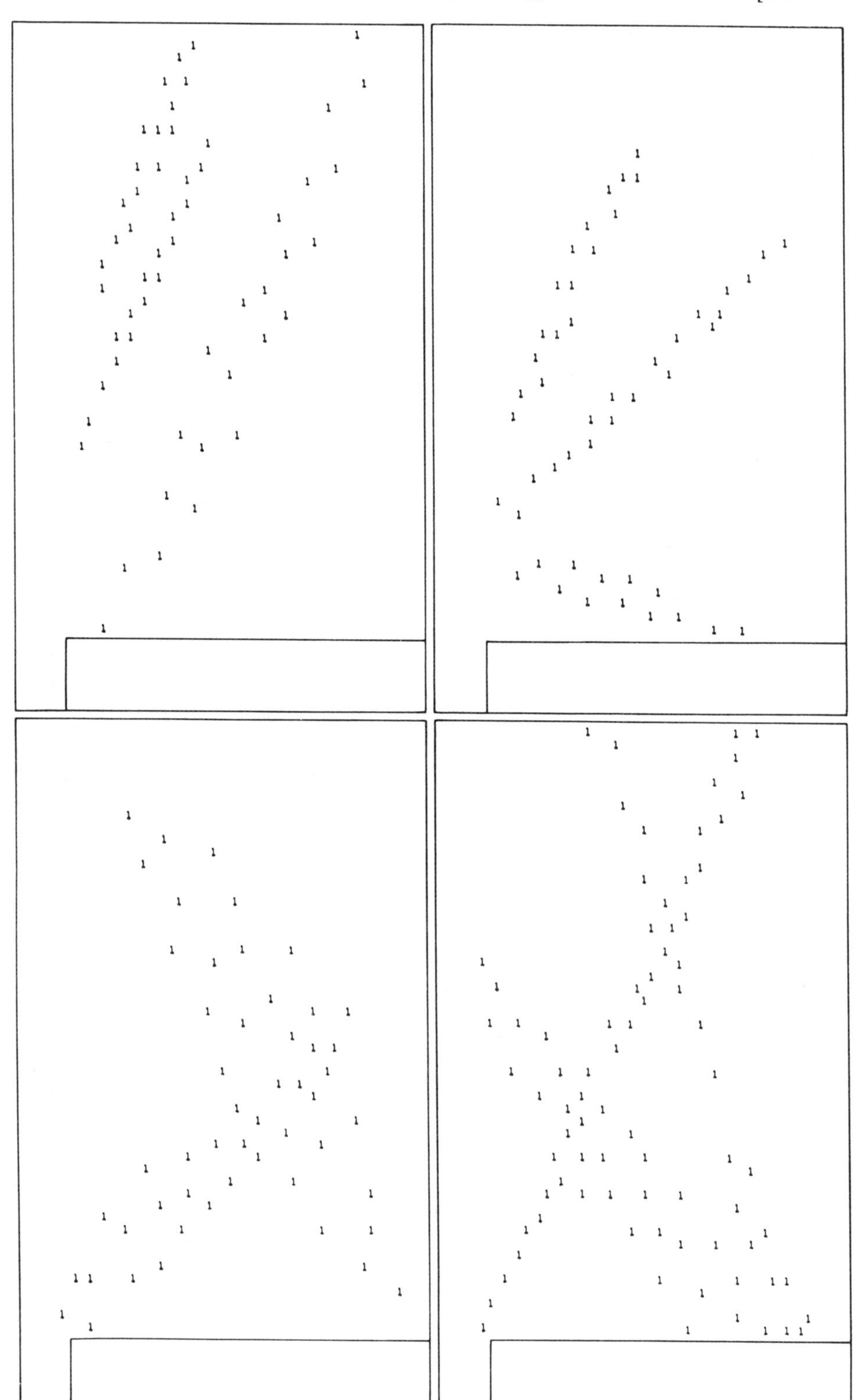

*Fig. 11.7*

will be more interesting to see how DETEXM and DWBEXM perform with the more practically oriented examples 2 and 4, and how DETEXM behaves with example 5 and DWBEXM behaves with examples 5 to 8 when the 'wrong' number of classes has been specified.

## 11.7 Sample Main Program for the Determinant Criterion and Results for the Test Examples

The sample main program for DETEXM shown in Table 11.11 (as well as that for DWBEXM) is almost identical to the program for TRWEXM listed in Table 11.5. There are differences, especially in the DIMENSION statements. With the determinant criterion, no individual measures of homogeneity (E(J), J=1, N) are output.

An important difference is that the given data matrix is transformed before being passed to DETEXM, to ensure that both columns have a mean of zero and variance of one. This is done with a call to TRAFOR (see Section 9.1) after the statements labelled 2. Because of the invariance property, the end partitions produced by DETEXM (and DWBEXM) are unaffected, in contrast to TRWEXM. We can therefore display the results, using the untransformed data matrix again. The intention of the normalization is to reduce the rounding error introduced by frequent up- and downdating. When all numbers are of the same order of magnitude, namely in the interval $[-1, 1]$, we expect fewer cancellation errors. This procedure may not be important with our small examples, but it is definitely recommended for the data sets used in practice. Operations within TRAFOR should perhaps even be done in double precision.

Table 11.12 shows the results of applying DETEXM to the realistic example 2 and 4 and to the special example 5, using NRMAX = 20 random starting partitions. The values of the objective function are shown for end partitions with 2 to 6 classes. The reader may confirm the continuing validity of statements about the number of passes needed and about the distribution of values of the objective function, which were made in the context of TRWEXM.

In Fig. 11.8 we show end partitions produced by DETEXM from examples 1 to 3 which differ significantly from those produced by TRWEXM. By 'significantly' we mean that clusters differ in the assignment of more than a few (1 to 3) points. The results for example 2 with $n = 3$ and $n = 5$ are particularly remarkable. In terms of their practical significance these results may be as good if not better than those produced by TRWEXM (see Fig. 11.4). Of course, the assumption underlying such statements is that we are dealing with optimal partitions. Although this optimality cannot be proved, it is nevertheless very probable in view of the distribution of the values of the objective functions in Tables 11.7 and 11.12.

In Fig. 11.9, the result for example 4 and $n = 2, \ldots, 5$ are reproduced. For $n = 3, 4, 5$ one can clearly see that the ellipsoids imposed have the same size and axis orientation. The reader may consider whether he regards the results of DETEXM as geometrically superior to those produced by TRWEXM, especially for $n = 3$.

*Table 11.11*

```
C     SAMPLE MAIN PROGRAM FOR DETEXM
C
C     MDX=200       MAXIMAL NUMBER OF OBJECTS
C     S=2           NUMBER OF VARIABLES (REQUIRED FOR GRAPHICAL OUTPUT)
C     NDX=9         MAXIMAL NUMBER OF CLUSTERS
C     MO=1          MINIMAL NUMBER OF CLUSTER MEMBERS
C
      INTEGER    Z(200),X1(200),X2(200),KIT(9,20),MJ(9)
      DIMENSION X(200,2),XBAR(9,2),DNNR(9,20)
      DIMENSION W(2,2),WBAR(2,2),XI(2),F(2)
      KI=5
      KO=6
      BIG=1.E50
      R=.999
      KPRINT=1
      MO=1
      N1=2
      N2=6
      NRMAX=20
    1 READ(KI,20) M
      IF(M.LE.0) STOP
      READ(KI,20) (X1(I),I=1,M)
      READ(KI,20) (X2(I),I=1,M)
      DO 2 I=1,M
      X(I,1)=X1(I)
    2 X(I,2)=X2(I)
      CALL TRAFOR (X,200,M,2,IFLAG)
      IF(IFLAG.NE.0) GOTO 1
      DO 6 N=N1,N2
            DD=BIG
            IY=37519
            DO 5 NR=1,NRMAX
                  CALL RANDP (M,N,Z,IY)
                  CALL DETEXM (X,200,M,2,2,Z,MO,MJ,XBAR,9,N,
     *                         D,IT,IFLAG,W,WBAR,XI,F)
                  IF(IFLAG.EQ.0) GOTO 3
                  DNNR(N,NR)=0.
                  KIT(N,NR)= -IFLAG
                  GOTO 5
    3             DNNR(N,NR)=D
                  KIT(N,NR)=IT
                  IF(D.GE.DD*R) GOTO 5
                  IF(KPRINT.EQ.0) GOTO 4
                  CALL PRINT (M,X1,X2,Z,N)
                  WRITE(KO,22)
                  WRITE(KO,21) M,N,NR,IT,D
                  WRITE(KO,22) (J,MJ(J),J=1,N)
    4             DD=D
    5       CONTINUE
    6 CONTINUE
      WRITE(KO,23) (J,J=N1,N2)
      DO 7 NR=1,NRMAX
    7 WRITE(KO,24) NR,(KIT(J,NR),DNNR(J,NR),J=N1,N2)
      GOTO 1
   20 FORMAT(16I5)
   21 FORMAT('0',17X,'M=',I3,'  N=',I1,'  NR=',I2,'  IT=',
     *       I2,'  D=',E11.4,'  (DETW)')
   22 FORMAT('0',17X,2(I1,' : (',I2,')  ',16X)/
     *           (18X,2(I1,' : (',I2,')  ',16X)))
   23 FORMAT('1',17X,I1,8I14)
   24 FORMAT(' ',I5,9(I3,E11.4))
      END
```

*Table 11.12*

| | 2 | 3 | 4 | 5 | 6 |
|---|---|---|---|---|---|
| 1 | 2 0.3034E+00 | 3 0.1326E+00 | 6 0.5610E-01 | 8 0.3764E-01 | 6 0.2757E-01 |
| 2 | 3 0.2599E+00 | 3 0.1426E+00 | 7 0.6142E-01 | 5 0.5237E-01 | 4 0.4316E-01 |
| 3 | 6 0.2599E+00 | 2 0.1426E+00 | 6 0.8712E-01 | 5 0.5237E-01 | 4 0.2681E-01 |
| 4 | 6 0.2599E+00 | 2 0.1326E+00 | 5 0.6260E-01 | 6 0.3764E-01 | 3 0.3027E-01 |
| 5 | 3 0.2394E+00 | 2 0.1407E+00 | 4 0.5610E-01 | 8 0.3962E-01 | 3 0.3635E-01 |
| 6 | 3 0.2599E+00 | 3 0.1644E+00 | 4 0.5610E-01 | 7 0.3764E-01 | 4 0.2761E-01 |
| 7 | 2 0.3034E+00 | 2 0.1426E+00 | 5 0.5610E-01 | 3 0.3764E-01 | 5 0.3400E-01 |
| 8 | 3 0.2599E+00 | 2 0.1407E+00 | 4 0.6495E-01 | 5 0.3962E-01 | 5 0.2601E-01 |
| 9 | 4 0.2394E+00 | 3 0.1479E+00 | 3 0.5610E-01 | 3 0.7256E-01 | 4 0.3068E-01 |
| 10 | 2 0.3034E+00 | 2 0.1326E+00 | 3 0.5610E-01 | 8 0.3700E-01 | 5 0.2614E-01 |
| 11 | 2 0.2604E+00 | 5 0.1041E+00 | 2 0.5610E-01 | 7 0.4198E-01 | 3 0.3265E-01 |
| 12 | 2 0.2604E+00 | 5 0.1041E+00 | 5 0.6260E-01 | 4 0.4185E-01 | 2 0.2925E-01 |
| 13 | 2 0.2394E+00 | 3 0.1562E+00 | 3 0.5610E-01 | 2 0.4036E-01 | 3 0.2681E-01 |
| 14 | 4 0.2599E+00 | 3 0.1326E+00 | 5 0.8712E-01 | 6 0.3764E-01 | 3 0.2925E-01 |
| 15 | 3 0.2599E+00 | 3 0.1479E+00 | 6 0.5610E-01 | 5 0.3764E-01 | 4 0.2925E-01 |
| 16 | 2 0.2394E+00 | 4 0.1426E+00 | 4 0.6838E-01 | 3 0.3764E-01 | 4 0.2755E-01 |
| 17 | 2 0.2394E+00 | 6 0.1041E+00 | 3 0.6142E-01 | 3 0.4185E-01 | 5 0.2761E-01 |
| 18 | 3 0.2394E+00 | 6 0.1041E+00 | 3 0.5610E-01 | 8 0.3764E-01 | 8 0.3248E-01 |
| 19 | 2 0.2604E+00 | 5 0.1041E+00 | 5 0.5610E-01 | 7 0.3962E-01 | 2 0.2681E-01 |
| 20 | 5 0.2599E+00 | 3 0.1426E+00 | 6 0.5610E-01 | 3 0.3764E-01 | 3 0.2883E-01 |

| | 2 | 3 | 4 | 5 | 6 |
|---|---|---|---|---|---|
| 1 | 2 0.2184E+00 | 3 0.8214E-01 | 3 0.4748E-01 | 7 0.1760E-01 | 6 0.1085E-01 |
| 2 | 4 0.2184E+00 | 2 0.1085E+00 | 4 0.3104E-01 | 4 0.1760E-01 | 6 0.1085E-01 |
| 3 | 4 0.2184E+00 | 3 0.8214E-01 | 4 0.4748E-01 | 7 0.1816E-01 | 4 0.1223E-01 |
| 4 | 4 0.2184E+00 | 2 0.1085E+00 | 4 0.3104E-01 | 5 0.1760E-01 | 9 0.1085E-01 |
| 5 | 2 0.1800E+00 | 8 0.1043E+00 | 7 0.3568E-01 | 5 0.2926E-01 | 12 0.2758E-01 |
| 6 | 4 0.2184E+00 | 5 0.1085E+00 | 6 0.3568E-01 | 6 0.2926E-01 | 11 0.1085E-01 |
| 7 | 2 0.2180E+00 | 3 0.8214E-01 | 3 0.4748E-01 | 4 0.3249E-01 | 3 0.1223E-01 |
| 8 | 4 0.2184E+00 | 2 0.1085E+00 | 7 0.3213E-01 | 5 0.1559E-01 | 5 0.1085E-01 |
| 9 | 4 0.2184E+00 | 5 0.1085E+00 | 5 0.4748E-01 | 10 0.3249E-01 | 5 0.2758E-01 |
| 10 | 5 0.2184E+00 | 3 0.8214E-01 | 4 0.3104E-01 | 6 0.1760E-01 | 6 0.1085E-01 |
| 11 | 7 0.2184E+00 | 2 0.8214E-01 | 3 0.4748E-01 | 8 0.3039E-01 | 6 0.1085E-01 |
| 12 | 2 0.1823E+00 | 3 0.8214E-01 | 6 0.3568E-01 | 6 0.3216E-01 | 7 0.1085E-01 |
| 13 | 4 0.2184E+00 | 3 0.8214E-01 | 7 0.3568E-01 | 6 0.1760E-01 | 6 0.8820E-02 |
| 14 | 4 0.2184E+00 | 3 0.1085E+00 | 4 0.4748E-01 | 6 0.3697E-01 | 12 0.2758E-01 |
| 15 | 2 0.1800E+00 | 8 0.1043E+00 | 7 0.3568E-01 | 7 0.2926E-01 | 5 0.1260E-01 |
| 16 | 2 0.2184E+00 | 5 0.1085E+00 | 4 0.4748E-01 | 7 0.2926E-01 | 4 0.3001E-01 |
| 17 | 4 0.2184E+00 | 2 0.1085E+00 | 3 0.4748E-01 | 7 0.3249E-01 | 8 0.2758E-01 |
| 18 | 3 0.2184E+00 | 2 0.1085E+00 | 6 0.3568E-01 | 7 0.1816E-01 | 15 0.1517E-01 |
| 19 | 2 0.1823E+00 | 6 0.7487E-01 | 3 0.3104E-01 | 4 0.2361E-01 | 6 0.1085E-01 |
| 20 | 5 0.2184E+00 | 3 0.1085E+00 | 8 0.3568E-01 | 5 0.3249E-01 | 6 0.3001E-01 |

| | 2 | 3 | 4 | 5 | 6 |
|---|---|---|---|---|---|
| 1 | 2 0.8908E-01 | 2 0.4050E-01 | 8 0.1053E-01 | 7 0.5159E-02 | 7 0.3599E-02 |
| 2 | 7 0.8908E-01 | 3 0.4050E-01 | 5 0.1053E-01 | 9 0.5159E-02 | 8 0.3599E-02 |
| 3 | 2 0.8908E-01 | 2 0.4050E-01 | 3 0.3085E-01 | 4 0.2552E-01 | 4 0.4546E-02 |
| 4 | 1 0.8908E-01 | 2 0.4050E-01 | 5 0.1053E-01 | 8 0.5010E-01 | 8 0.3599E-02 |
| 5 | 1 0.8908E-01 | 4 0.1251E+00 | 4 0.1053E-01 | 8 0.5010E-01 | 5 0.5038E-02 |
| 6 | 8 0.8908E-01 | 3 0.4050E-01 | 4 0.1053E-01 | 7 0.5159E-02 | 8 0.3599E-02 |
| 7 | 3 0.8908E-01 | 2 0.4050E-01 | 6 0.1053E-01 | 4 0.5731E-02 | 5 0.4546E-02 |
| 8 | 5 0.8908E-01 | 7 0.1274E+00 | 7 0.1053E-01 | 7 0.5159E-02 | 7 0.3599E-02 |
| 9 | 4 0.8908E-01 | 4 0.2502E-01 | 4 0.1053E-01 | 6 0.7910E-02 | 7 0.3636E-02 |
| 10 | 1 0.8908E-01 | 2 0.4050E-01 | 4 0.1053E-01 | 8 0.5159E-02 | 4 0.5038E-02 |
| 11 | 2 0.8908E-01 | 2 0.4050E-01 | 6 0.1053E-01 | 8 0.5159E-02 | 6 0.3636E-02 |
| 12 | 3 0.8908E-01 | 6 0.2502E-01 | 4 0.1053E-01 | 7 0.5159E-02 | 6 0.4941E-02 |
| 13 | 2 0.8908E-01 | 2 0.4050E-01 | 3 0.3085E-01 | 8 0.5159E-02 | 5 0.3599E-02 |
| 14 | 2 0.8908E-01 | 2 0.4050E-01 | 3 0.3085E-01 | 4 0.2552E-01 | 7 0.1034E-01 |
| 15 | 7 0.8908E-01 | 7 0.1274E+00 | 5 0.1053E-01 | 7 0.5159E-02 | 8 0.3599E-02 |
| 16 | 2 0.8908E-01 | 2 0.4050E-01 | 3 0.3249E-01 | 4 0.2552E-01 | 5 0.4546E-02 |
| 17 | 7 0.8908E-01 | 2 0.4050E-01 | 5 0.7766E-01 | 10 0.5159E-02 | 5 0.4546E-02 |
| 18 | 1 0.8908E-01 | 2 0.4050E-01 | 3 0.3085E-01 | 6 0.5159E-02 | 6 0.3599E-02 |
| 19 | 1 0.8908E-01 | 2 0.4050E-01 | 4 0.1053E-01 | 4 0.2552E-01 | 9 0.1034E-01 |
| 20 | 3 0.8908E-01 | 2 0.4050E-01 | 5 0.1053E-01 | 6 0.5159E-02 | 4 0.4720E-02 |

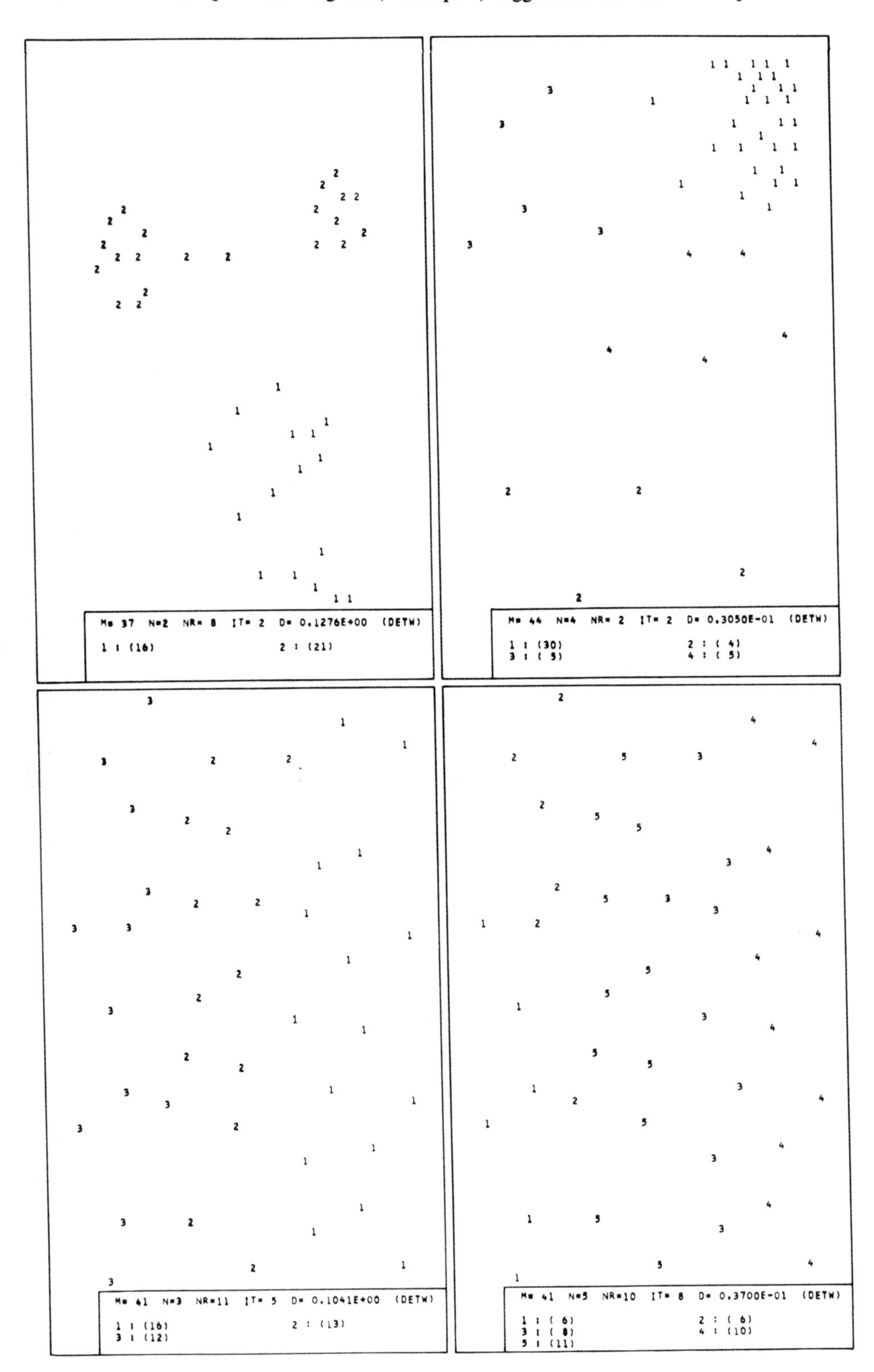
M= 37 N=2 NR= 8 IT= 2 D= 0.1276E+00 (DETW)
1 : (16) 2 : (21)
M= 44 N=4 NR= 2 IT= 2 D= 0.3050E-01 (DETW)
1 : (30) 2 : ( 4)
3 : ( 5) 4 : ( 5)
M= 41 N=3 NR=11 IT= 5 D= 0.1041E+00 (DETW)
1 : (16) 2 : (13)
3 : (12)
M= 41 N=5 NR=10 IT= 8 D= 0.3700E-01 (DETW)
1 : ( 6) 2 : ( 6)
3 : ( 8) 4 : (10)
5 : (11)

*Fig. 11.8*

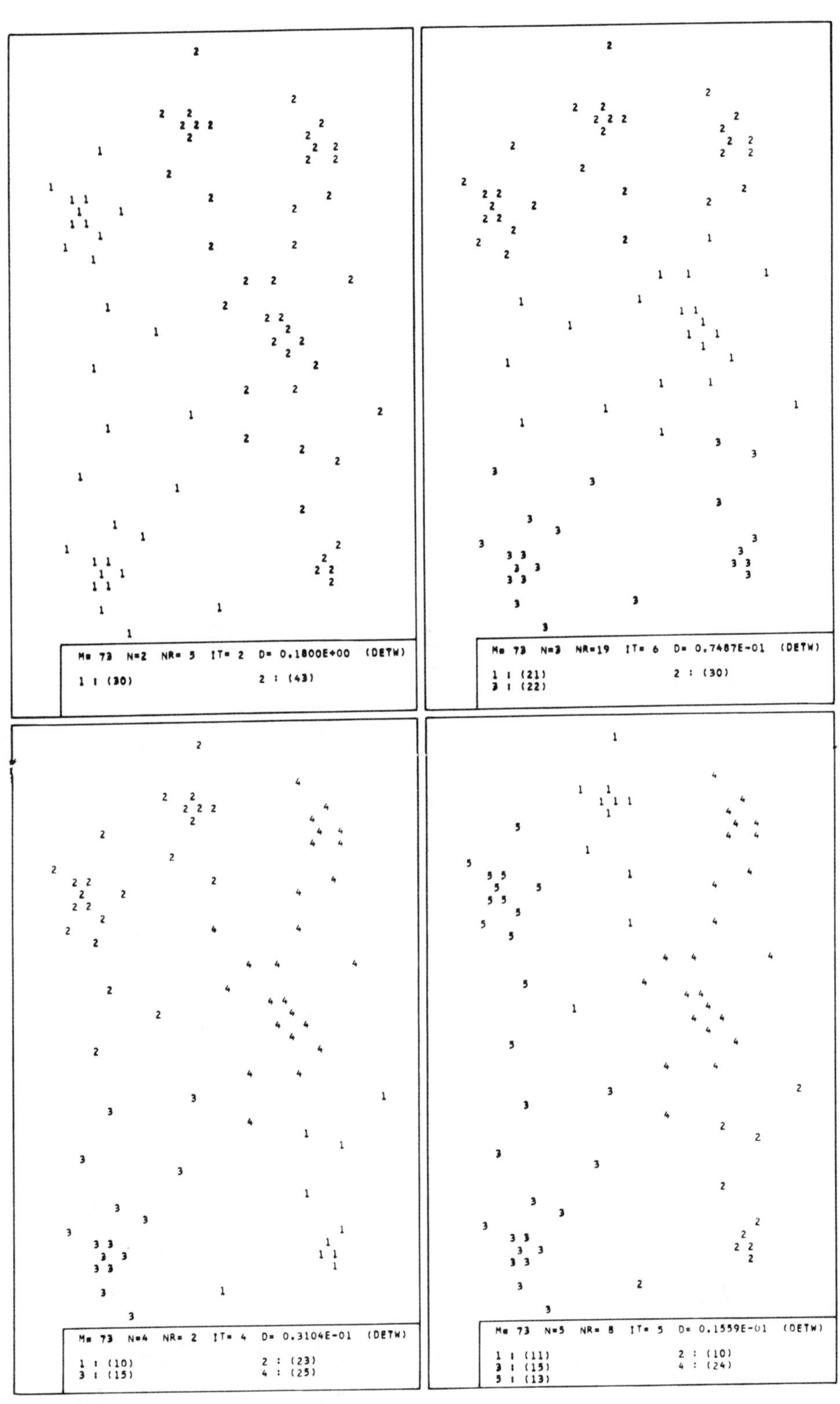
M= 73 N=2 NR= 5 IT= 2 D= 0.1800E+00 (DETW)
1 : (30) 2 : (43)
M= 73 N=3 NR=19 IT= 6 D= 0.7487E-01 (DETW)
1 : (21) 2 : (30)
3 : (22)
M= 73 N=4 NR= 2 IT= 4 D= 0.3104E-01 (DETW)
1 : (10) 2 : (23)
3 : (15) 4 : (25)
M= 73 N=5 NR= 8 IT= 5 D= 0.1559E-01 (DETW)
1 : (11) 2 : (10)
3 : (15) 4 : (24)
5 : (13)

*Fig. 11.9*

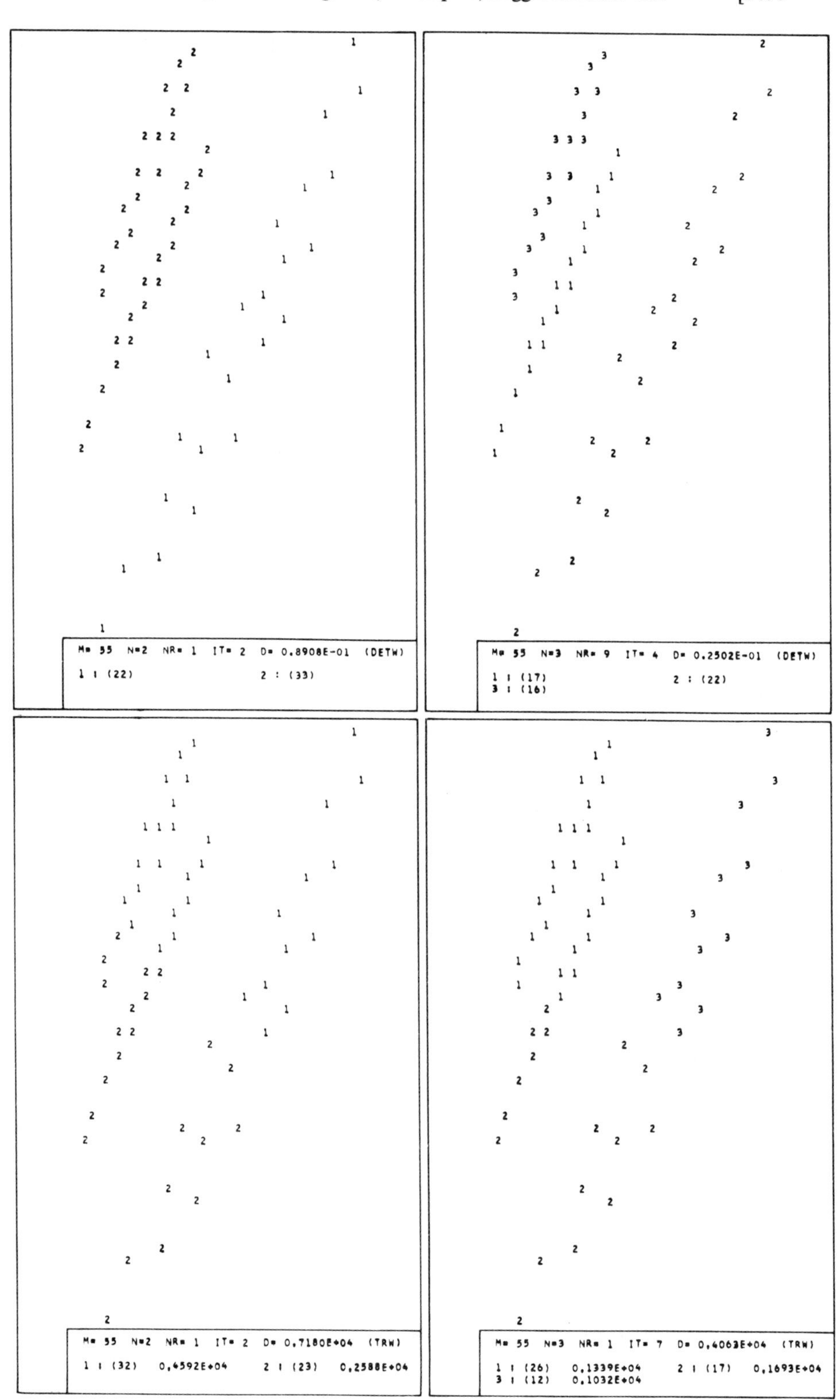
M= 55 N=2 NR= 1 IT= 2 D= 0.8908E-01 (DETW)
1 : (22) 2 : (33)
M= 55 N=3 NR= 9 IT= 4 D= 0.2502E-01 (DETW)
1 : (17) 2 : (22)
3 : (16)
M= 55 N=2 NR= 1 IT= 2 D= 0,7180E+04 (TRW)
1 : (32) 0,4592E+04 2 : (23) 0,2588E+04
M= 55 N=3 NR= 1 IT= 7 D= 0,4063E+04 (TRW)
1 : (26) 0,1339E+04 2 : (17) 0,1693E+04
3 : (12) 0,1032E+04

*Fig. 11.10*

For example 5, the results using TRWEXM and DETEXM are juxtaposed for $n = 2$ and $n = 3$. Apparently, both subroutines perform in the way that the theory suggests. DETEXM finds the structure for $n = 2$ and $n = 3$ classes. The last block of numbers in Table 11.12 which relates to example 5, indicates that $n = 2$ is the 'correct' number of classes, since with $n = 2$ all 20 starting partitions lead to the same value of the objective function, whereas, for $n = 3$, the minimum is attained only with two starting partitions. With TRWEXM the situation is reversed: for $n = 3$, the minimum is attained 20 times, for $n = 2$ only 16 times.

The computer time required for a total of 12 examples, of which 8 have now been given explicitly, for $n = 2, \ldots, 6$ and NRMAX = 20 starting partitions each and KPRINT = 1 using DETEXM was 1522 CPU seconds on the TR 440. The corresponding time using TRWEXM was 295 seconds. Since each exchange attempt under TRWEXM takes O(mns) operations, compared with O($mns^2$) operations under DETEXM and DWBEXM, the ratio of computer time used will be even more unfavourable for $s > 2$.

## 11.8 Sample Main Program for the Criterion of Adaptive Distances, and Results for the Test Examples

There is only a slight difference between the main program for DWBEXM listed in Table 11.13 and the main program for DETEXM. To set $m_0 = s + 1$ and $\beta = 1/s$ for our two-dimensional examples, the two statements

```
MO = 3
BETA = .5
```

have been added. BETA = .5 is also indicated on the graphical output by 'DWB .5'. By merely changing the second statement to BETA = $\beta$ with some $\beta > 0$, the criterion (4.15) can be computed. The homogeneities of individual clusters are again output in (E(J), J=1, N). For the same reason as in the last section. TRAFOR is used again. However, with large values of $\beta$ this does not prevent the sensitivity to rounding errors mentioned previously.

Table 11.14 and 11.15 again contain values for end partitions with $n = 2, \ldots, 6$ classes for 20 starting partitions each. Table 11.14 relates to the second and fourth standard examples. Table 11.15 to examples 6, 7 and 8 which are specifically tailored to the criterion of adaptive distances. Previous statements about the number of passes and the distribution of the values of the objective function generated by the exchange method continue to hold. However, with larger numbers of classes the error condition IFLAG = 7 occurs relatively frequently. This means that in UPDATE, probably during downdating, the determinant of a scatter matrix become smaller than EPS3 = 1.E-6 for at least one of the classes, and that the process was aborted. It is worth noting that this effect does not occur with the 'correct' number of classes. Occasionally, the errors IFLAG = 2 and IFLAG = 6 occur. IFLAG = 2 means that there was a class with fewer than M0 = 3 elements in the starting partition. IFLAG = 6

*Table 11.13*

```
C   SAMPLE MAIN PROGRAM FOR DWBEXM
C
C   MDX=200      MAXIMAL NUMBER OF OBJECTS
C   S=2          NUMBER OF VARIABLES (REQUIRED FOR GRAPHICAL OUTPUT)
C   NDX=9        MAXIMAL NUMBER OF CLUSTERS
C   MO=3         MINIMAL NUMBER OF CLUSTER MEMBERS
C
      INTEGER   Z(200),X1(200),X2(200),KIT(9,20),MJ(9)
      DIMENSION X(200,2),XBAR(9,2),E(9),DNNR(9,20)
      DIMENSION W(9,2,2),WJ(2,2),WP(2,2),WQ(2,2),F(2)
      KI=5
      KO=6
      BIG=1.E50
      R=.999
      KPRINT=1
      MO=3
      BETA=.5
      N1=2
      N2=6
      NRMAX=20
    1 READ(KI,20) M
      IF(M.LE.0) STOP
      READ(KI,20) (X1(I),I=1,M)
      READ(KI,20) (X2(I),I=1,M)
      DO 2 I=1,M
      X(I,1)=X1(I)
    2 X(I,2)=X2(I)
      CALL TRAFOR (X,200,M,2,IFLAG)
      IF(IFLAG.NE.0) GOTO 1
      DO 6 N=N1,N2
            DD=BIG
            IY=37519
            DO 5 NR=1,NRMAX
                  CALL RANDP (M,N,Z,IY)
                  CALL DWBEXM (X,200,M,2,2,Z,MO,MJ,XBAR,9,N,
     *                         BETA,E,D,IT,IFLAG,W,WJ,WP,WQ,F)
                  IF(IFLAG.EQ.0) GOTO 3
                  DNNR(N,NR)=0.
                  KIT(N,NR)= -IFLAG
                  GOTO 5
    3             DNNR(N,NR)=D
                  KIT(N,NR)=IT
                  IF(D.GE.DD*R) GOTO 5
                  IF(KPRINT.EQ.0) GOTO 4
                  CALL PRINT (M,X1,X2,Z,N)
                  WRITE(KO,22)
                  WRITE(KO,21) M,N,NR,IT,D
                  WRITE(KO,22) (J,MJ(J),E(J),J=1,N)
    4             DD=D
    5       CONTINUE
    6 CONTINUE
      WRITE(KO,23) (J,J=N1,N2)
      DO 7 NR=1,NRMAX
    7 WRITE(KO,24) NR,(KIT(J,NR),DNNR(J,NR),J=N1,N2)
      GOTO 1
   20 FORMAT(16I5)
   21 FORMAT('0',17X,'M=',I3,'  N=',I1,'  NR=',I2,'  IT=',
     *         I2,'  D=',E11.4,' (DWB .5)')
   22 FORMAT('0',17X,2(I1,' : (',I2,')  ',E11.4,5X)/
     *              (18X,2(I1,' : (',I2,')  ',E11.4,5X)))
   23 FORMAT('1',17X,I1,8I14)
   24 FORMAT(' ',I5,9(I3,E11.4))
      END
```

also indicates a collapse of the up- and downdating process for reasons which we have not investigated further. It is, however, plausible that these errors occur more frequently here than with DETEXM, since changes are here made to the individual scatter matrices $W_j$ consisting of fewer dyadic products, rather than to the sum of the $W_j$.

*Table 11.14*

| | 2 | | 3 | | 4 | | 5 | | 6 | |
|---|---|---|---|---|---|---|---|---|---|---|
| 1 | 3 | 0.4839E+00 | 2 | 0,3362E+00 | 3 | 0,2440E+00 | 3 | 0.1846E+00 | -7 | 0.0 |
| 2 | 3 | 0.5089E+00 | 3 | 0,3391E+00 | 3 | 0,2441E+00 | 3 | 0.1869E+00 | 2 | 0.1704E+00 |
| 3 | 3 | 0.4839E+00 | 7 | 0,3389E+00 | 2 | 0,2347E+00 | 3 | 0.1882E+00 | 2 | 0.1541E+00 |
| 4 | 2 | 0.5089E+00 | 2 | 0,3362E+00 | 3 | 0,2397E+00 | -7 | 0.0 | 4 | 0.1855E+00 |
| 5 | 2 | 0.4839E+00 | 3 | 0,3362E+00 | 2 | 0,2397E+00 | 3 | 0.1871E+00 | -7 | 0.0 |
| 6 | 4 | 0.4839E+00 | 3 | 0,3333E+00 | 2 | 0,2347E+00 | 3 | 0.1953E+00 | 2 | 0.1475E+00 |
| 7 | 4 | 0.6342E+00 | 4 | 0,3333E+00 | 3 | 0,2401E+00 | 4 | 0.1887E+00 | -2 | 0.0 |
| 8 | 3 | 0.4897E+00 | 2 | 0,3362E+00 | 7 | 0,2348E+00 | 3 | 0.1790E+00 | -7 | 0.0 |
| 9 | 2 | 0.4839E+00 | 3 | 0,3391E+00 | 3 | 0,2347E+00 | 3 | 0,1941E+00 | -2 | 0.0 |
| 10 | 3 | 0.4839E+00 | 3 | 0,3369E+00 | 7 | 0,2334E+00 | 3 | 0.1767E+00 | -2 | 0.0 |
| 11 | 2 | 0.4897E+00 | 3 | 0,3391E+00 | 5 | 0,2349E+00 | -7 | 0.0 | 3 | 0.1491E+00 |
| 12 | 3 | 0.5089E+00 | 4 | 0,3338E+00 | 3 | 0,2404E+00 | 5 | 0.1780E+00 | 2 | 0.1444E+00 |
| 13 | 3 | 0.4839E+00 | 2 | 0,3343E+00 | 4 | 0,2381E+00 | 5 | 0.1780E+00 | -7 | 0.0 |
| 14 | 4 | 0.5089E+00 | 2 | 0,3339E+00 | -7 | 0.0 | 2 | 0.1973E+00 | -7 | 0.0 |
| 15 | 3 | 0.5089E+00 | 4 | 0,3262E+00 | 7 | 0,2270E+00 | 3 | 0.1881E+00 | 4 | 0.1469E+00 |
| 16 | 2 | 0.4839E+00 | 3 | 0,3262E+00 | 3 | 0,2554E+00 | 3 | 0.1869E+00 | -2 | 0.0 |
| 17 | 2 | 0.4839E+00 | 3 | 0,3343E+00 | 7 | 0,2220E+00 | 2 | 0.1843E+00 | -7 | 0.0 |
| 18 | 2 | 0.4839E+00 | 3 | 0,3391E+00 | 4 | 0,2334E+00 | 3 | 0.2007E+00 | -7 | 0.0 |
| 19 | 3 | 0.4839E+00 | 2 | 0,3338E+00 | 4 | 0,2537E+00 | 4 | 0.1790E+00 | 3 | 0.1356E+00 |
| 20 | 2 | 0.5089E+00 | 3 | 0,3338E+00 | 5 | 0,2339E+00 | 7 | 0.1790E+00 | 3 | 0.1439E+00 |

| | 2 | | 3 | | 4 | | 5 | | 6 | |
|---|---|---|---|---|---|---|---|---|---|---|
| 1 | 4 | 0.4584E+00 | 2 | 0,2827E+00 | 4 | 0.1548E+00 | -7 | 0.0 | 5 | 0.8308E-01 |
| 2 | 3 | 0.4217E+00 | 2 | 0,2645E+00 | 3 | 0,1870E+00 | -7 | 0.0 | 5 | 0.8477E-01 |
| 3 | 1 | 0.4217E+00 | 3 | 0,2476E+00 | 4 | 0.2282E+00 | 2 | 0.1236E+00 | 4 | 0.8788E-01 |
| 4 | 4 | 0.4584E+00 | 5 | 0,2356E+00 | 2 | 0.1806E+00 | 2 | 0.1271E+00 | -7 | 0.0 |
| 5 | 2 | 0.4102E+00 | 2 | 0,2801E+00 | 4 | 0.1576E+00 | 6 | 0.1257E+00 | 3 | 0.8683E-01 |
| 6 | 5 | 0.4584E+00 | 3 | 0,2476E+00 | 7 | 0.1650E+00 | 4 | 0.1202E+00 | 5 | 0.1115E+00 |
| 7 | 3 | 0.4217E+00 | 2 | 0,2476E+00 | -7 | 0.0 | -7 | 0.0 | 4 | 0.8007E-01 |
| 8 | 2 | 0.4217E+00 | 6 | 0,2399E+00 | 2 | 0.1565E+00 | 4 | 0.1330E+00 | -7 | 0.0 |
| 9 | 1 | 0.4217E+00 | 2 | 0,2827E+00 | 2 | 0.1565E+00 | 3 | 0.1080E+00 | 8 | 0.9341E-01 |
| 10 | 2 | 0.4584E+00 | 3 | 0,2476E+00 | 4 | 0.1907E+00 | 7 | 0.1259E+00 | 7 | 0.8177E-01 |
| 11 | 6 | 0.5477E+00 | 5 | 0,2684E+00 | 4 | 0.1855E+00 | 3 | 0.1137E+00 | 4 | 0.8007E-01 |
| 12 | 2 | 0.4584E+00 | 3 | 0,2476E+00 | 2 | 0.1855E+00 | 4 | 0.1400E+00 | -7 | 0.0 |
| 13 | 2 | 0.4618E+00 | 2 | 0,2827E+00 | 3 | 0,1870E+00 | 7 | 0.1400E+00 | -7 | 0.0 |
| 14 | 1 | 0.4217E+00 | 2 | 0,2827E+00 | 3 | 0.1590E+00 | 3 | 0.1249E+00 | -7 | 0.0 |
| 15 | 2 | 0.4217E+00 | 2 | 0,2476E+00 | 4 | 0.1806E+00 | -7 | 0.0 | -7 | 0.0 |
| 16 | 2 | 0.4217E+00 | 2 | 0,2827E+00 | 5 | 0,1907E+00 | 3 | 0.1258E+00 | 3 | 0.8702E-01 |
| 17 | 4 | 0.4584E+00 | 4 | 0,3109E+00 | 8 | 0.1574E+00 | 6 | 0.1457E+00 | -7 | 0.0 |
| 18 | 2 | 0.5601E+00 | 4 | 0,2476E+00 | 2 | 0.1590E+00 | 3 | 0.1138E+00 | 4 | 0.1132E+00 |
| 19 | 3 | 0.4217E+00 | 4 | 0,2476E+00 | 3 | 0.1692E+00 | 4 | 0.1301E+00 | 3 | 0.8690E-01 |
| 20 | 5 | 0.4254E+00 | 2 | 0,2827E+00 | 5 | 0.1956E+00 | 3 | 0.1137E+00 | -7 | 0.0 |

Applied to the first example, DWBEXM shows nothing new, as expected. For examples 2 and 3, Fig. 11.11 displays the best of the twenty end partitions for $n = 3$ and $n = 4$ classes. Compare this with previous results of TRWEXM and DETEXM in Figs. 11.3, 11.4 and 11.8. Fig. 11.12 comprises the best end partitions for example 4 and $n = 3, 4, 5$ as well as the best end partition for example 5 and $n = 3$ classes. The latter differs considerably from the corresponding one produced by DETEXM.

*Table 11.15*

| | | 2 | | 3 | | 4 | | 5 | | 6 |
|---|---|---|---|---|---|---|---|---|---|---|
| 1 | 3 | 0.4701E+00 | 2 | 0.8155E-01 | 4 | 0.6448E-01 | -7 | 0.0 | -7 | 0.0 |
| 2 | 2 | 0.4701E+00 | 3 | 0.3084E+00 | 4 | 0.6448E-01 | -7 | 0.0 | 3 | 0.5099E-01 |
| 3 | 3 | 0.4365E+00 | 3 | 0.2334E+00 | 3 | 0.2073E+00 | -7 | 0.0 | -7 | 0.0 |
| 4 | 3 | 0.4701E+00 | 2 | 0.8155E-01 | 3 | 0.6448E-01 | -7 | 0.0 | -7 | 0.0 |
| 5 | 2 | 0.4368E+00 | 2 | 0.8155E-01 | 1 | 0.6695E-01 | 5 | 0.5158E-01 | -7 | 0.0 |
| 6 | 7 | 0.5603E+00 | 2 | 0.8155E-01 | 4 | 0.6448E-01 | 3 | 0.5158E-01 | -2 | 0.0 |
| 7 | 2 | 0.4701E+00 | 3 | 0.8155E-01 | 3 | 0.6448E-01 | 4 | 0.5196E-01 | -7 | 0.0 |
| 8 | 2 | 0.4365E+00 | 2 | 0.8155E-01 | 3 | 0.2073E+00 | -6 | 0.0 | -6 | 0.0 |
| 9 | 3 | 0.4701E+00 | 2 | 0.8155E-01 | 4 | 0.6448E-01 | -7 | 0.0 | -7 | 0.0 |
| 10 | 3 | 0.4701E+00 | 3 | 0.8155E-01 | 4 | 0.6448E-01 | -7 | 0.0 | -7 | 0.0 |
| 11 | 1 | 0.4701E+00 | 2 | 0.8155E-01 | 4 | 0.8075E-01 | -7 | 0.0 | -6 | 0.0 |
| 12 | 3 | 0.4701E+00 | 2 | 0.8155E-01 | 3 | 0.6448E-01 | -7 | 0.0 | -7 | 0.0 |
| 13 | 2 | 0.4365E+00 | 2 | 0.8155E-01 | 3 | 0.2073E+00 | -7 | 0.0 | -2 | 0.0 |
| 14 | 3 | 0.4701E+00 | 2 | 0.8155E-01 | 2 | 0.6695E-01 | -7 | 0.0 | -7 | 0.0 |
| 15 | 4 | 0.4413E+00 | 2 | 0.8155E-01 | 4 | 0.6448E-01 | -7 | 0.0 | -7 | 0.0 |
| 16 | 2 | 0.4701E+00 | 3 | 0.2334E+00 | 4 | 0.6448E-01 | -7 | 0.0 | -7 | 0.0 |
| 17 | 3 | 0.4701E+00 | 4 | 0.3325E+00 | -7 | 0.0 | -7 | 0.0 | 3 | 0.6382E-01 |
| 18 | 2 | 0.4365E+00 | 2 | 0.8155E-01 | 4 | 0.6448E-01 | -7 | 0.0 | -6 | 0.0 |
| 19 | 2 | 0.4365E+00 | 2 | 0.8155E-01 | -7 | 0.0 | -7 | 0.0 | -7 | 0.0 |
| 20 | 2 | 0.4701E+00 | 2 | 0.8155E-01 | 4 | 0.6506E-01 | -7 | 0.0 | -7 | 0.0 |

| | | 2 | | 3 | | 4 | | 5 | | 6 |
|---|---|---|---|---|---|---|---|---|---|---|
| 1 | 4 | 0.3147E+00 | 3 | 0.2500E+00 | 4 | 0.1467E+00 | -7 | 0.0 | 5 | 0.9381E-01 |
| 2 | 2 | 0.3139E+00 | 7 | 0.2001E+00 | 12 | 0.1367E+00 | -7 | 0.0 | 5 | 0.1131E+00 |
| 3 | 4 | 0.3147E+00 | 4 | 0.2117E+00 | 2 | 0.1827E+00 | 4 | 0.1351E+00 | 5 | 0.9336E-01 |
| 4 | 6 | 0.3147E+00 | 3 | 0.2500E+00 | 3 | 0.1566E+00 | 3 | 0.1421E+00 | 3 | 0.1262E+00 |
| 5 | 6 | 0.3147E+00 | 4 | 0.2500E+00 | 5 | 0.1467E+00 | 2 | 0.1404E+00 | 5 | 0.1106E+00 |
| 6 | 3 | 0.3147E+00 | 4 | 0.2500E+00 | 6 | 0.1566E+00 | 9 | 0.1023E+00 | -7 | 0.0 |
| 7 | 4 | 0.4600E+00 | 4 | 0.2505E+00 | 3 | 0.1827E+00 | -7 | 0.0 | -7 | 0.0 |
| 8 | 4 | 0.3147E+00 | 3 | 0.2500E+00 | -7 | 0.0 | 6 | 0.1248E+00 | 4 | 0.1294E+00 |
| 9 | 7 | 0.3147E+00 | 3 | 0.2500E+00 | 4 | 0.1467E+00 | -7 | 0.0 | 3 | 0.9012E-01 |
| 10 | 5 | 0.3147E+00 | 5 | 0.2468E+00 | 2 | 0.1853E+00 | -7 | 0.0 | -7 | 0.0 |
| 11 | 6 | 0.3147E+00 | 7 | 0.2001E+00 | 3 | 0.1467E+00 | 4 | 0.1242E+00 | -7 | 0.0 |
| 12 | 6 | 0.3147E+00 | 3 | 0.2500E+00 | 2 | 0.2007E+00 | 3 | 0.1522E+00 | -7 | 0.0 |
| 13 | 5 | 0.3147E+00 | 9 | 0.2001E+00 | 2 | 0.2007E+00 | 3 | 0.1122E+00 | 3 | 0.1079E+00 |
| 14 | 5 | 0.3147E+00 | 4 | 0.2500E+00 | -7 | 0.0 | 4 | 0.1154E+00 | 3 | 0.1176E+00 |
| 15 | 5 | 0.3147E+00 | 3 | 0.2500E+00 | 4 | 0.1809E+00 | 5 | 0.1153E+00 | 4 | 0.1084E+00 |
| 16 | 6 | 0.3147E+00 | 3 | 0.2500E+00 | 4 | 0.1467E+00 | -7 | 0.0 | 4 | 0.9384E-01 |
| 17 | 6 | 0.3147E+00 | 4 | 0.2500E+00 | 4 | 0.1593E+00 | -6 | 0.0 | -6 | 0.0 |
| 18 | 6 | 0.3147E+00 | 11 | 0.2001E+00 | 2 | 0.1467E+00 | 4 | 0.1238E+00 | 4 | 0.1035E+00 |
| 19 | 5 | 0.3147E+00 | 4 | 0.2500E+00 | 3 | 0.1799E+00 | 3 | 0.1404E+00 | -7 | 0.0 |
| 20 | 6 | 0.3147E+00 | 4 | 0.2500E+00 | 4 | 0.1779E+00 | -6 | 0.0 | -7 | 0.0 |

| | | 2 | | 3 | | 4 | | 5 | | 6 |
|---|---|---|---|---|---|---|---|---|---|---|
| 1 | 2 | 0.4320E+00 | 2 | 0.2587E+00 | 4 | 0.1998E+00 | 6 | 0.9290E-01 | -7 | 0.0 |
| 2 | 2 | 0.4320E+00 | 5 | 0.1437E+00 | 3 | 0.1902E+00 | 6 | 0.9290E-01 | 5 | 0.7718E-01 |
| 3 | 2 | 0.4320E+00 | 4 | 0.2394E+00 | 4 | 0.1080E+00 | 6 | 0.8752E-01 | -6 | 0.0 |
| 4 | 3 | 0.4320E+00 | 3 | 0.2586E+00 | 13 | 0.1703E+00 | 11 | 0.8538E-01 | 3 | 0.1182E+00 |
| 5 | 2 | 0.4320E+00 | 2 | 0.2574E+00 | 6 | 0.1194E+00 | 5 | 0.1117E+00 | 4 | 0.1111E+00 |
| 6 | 2 | 0.4320E+00 | 2 | 0.2587E+00 | 4 | 0.1919E+00 | -7 | 0.0 | -7 | 0.0 |
| 7 | 2 | 0.4320E+00 | 3 | 0.2586E+00 | 5 | 0.1077E+00 | 5 | 0.1017E+00 | 4 | 0.8166E-01 |
| 8 | 3 | 0.4320E+00 | 8 | 0.2484E+00 | 5 | 0.1844E+00 | 4 | 0.1296E+00 | 8 | 0.9874E-01 |
| 9 | 2 | 0.4320E+00 | 4 | 0.2394E+00 | 6 | 0.1077E+00 | -7 | 0.0 | -7 | 0.0 |
| 10 | 2 | 0.4320E+00 | 3 | 0.2586E+00 | 3 | 0.2128E+00 | 6 | 0.8752E-01 | -7 | 0.0 |
| 11 | 2 | 0.4699E+00 | 3 | 0.2586E+00 | 7 | 0.1077E+00 | 12 | 0.8752E-01 | -7 | 0.0 |
| 12 | 2 | 0.4320E+00 | 5 | 0.2394E+00 | 3 | 0.1988E+00 | 6 | 0.9290E-01 | 4 | 0.8558E-01 |
| 13 | 2 | 0.4320E+00 | 3 | 0.2573E+00 | 5 | 0.1077E+00 | 3 | 0.1343E+00 | -7 | 0.0 |
| 14 | 2 | 0.4320E+00 | 2 | 0.2587E+00 | 9 | 0.1077E+00 | 12 | 0.8752E-01 | -7 | 0.0 |
| 15 | 2 | 0.4674E+00 | 3 | 0.2586E+00 | 6 | 0.1077E+00 | -7 | 0.0 | 4 | 0.8237E-01 |
| 16 | 1 | 0.4320E+00 | 3 | 0.2586E+00 | 3 | 0.2128E+00 | 3 | 0.1341E+00 | 8 | 0.8379E-01 |
| 17 | 1 | 0.4320E+00 | 3 | 0.2394E+00 | 9 | 0.1077E+00 | 7 | 0.9312E-01 | 7 | 0.6564E-01 |
| 18 | 2 | 0.4320E+00 | 3 | 0.2586E+00 | 10 | 0.1703E+00 | 11 | 0.8752E-01 | 8 | 0.8379E-01 |
| 19 | 2 | 0.4320E+00 | 3 | 0.2586E+00 | 3 | 0.1902E+00 | 4 | 0.1632E+00 | -7 | 0.0 |
| 20 | 2 | 0.4320E+00 | 3 | 0.2587E+00 | 4 | 0.1919E+00 | 5 | 0.1267E+00 | 4 | 0.8166E-01 |

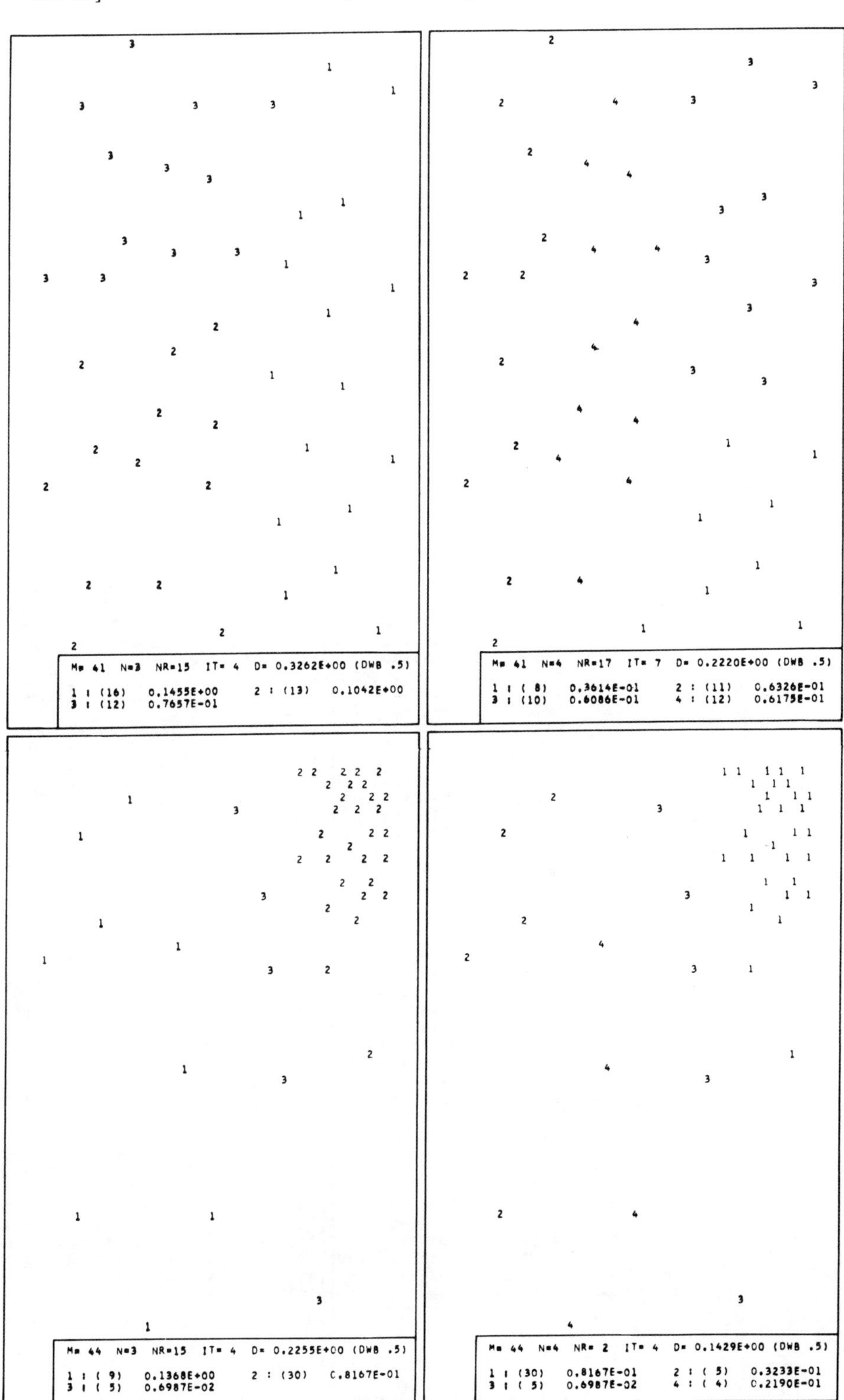
M= 41 N=3 NR=15 IT= 4 D= 0.3262E+00 (DWB .5)
1 : (16) 0.1455E+00 2 : (13) 0.1042E+00
3 : (12) 0.7657E-01
M= 41 N=4 NR=17 IT= 7 D= 0.2220E+00 (DWB .5)
1 : ( 8) 0.3614E-01 2 : (11) 0.6326E-01
3 : (10) 0.6086E-01 4 : (12) 0.6175E-01
M= 44 N=3 NR=15 IT= 4 D= 0.2255E+00 (DWB .5)
1 : ( 9) 0.1368E+00 2 : (30) C.8167E-01
3 : ( 5) 0.6987E-02
M= 44 N=4 NR= 2 IT= 4 D= 0.1429E+00 (DWB .5)
1 : (30) 0.8167E-01 2 : ( 5) 0.3233E-01
3 : ( 5) 0.6987E-02 4 : ( 4) 0.2190E-01

*Fig. 11.11*

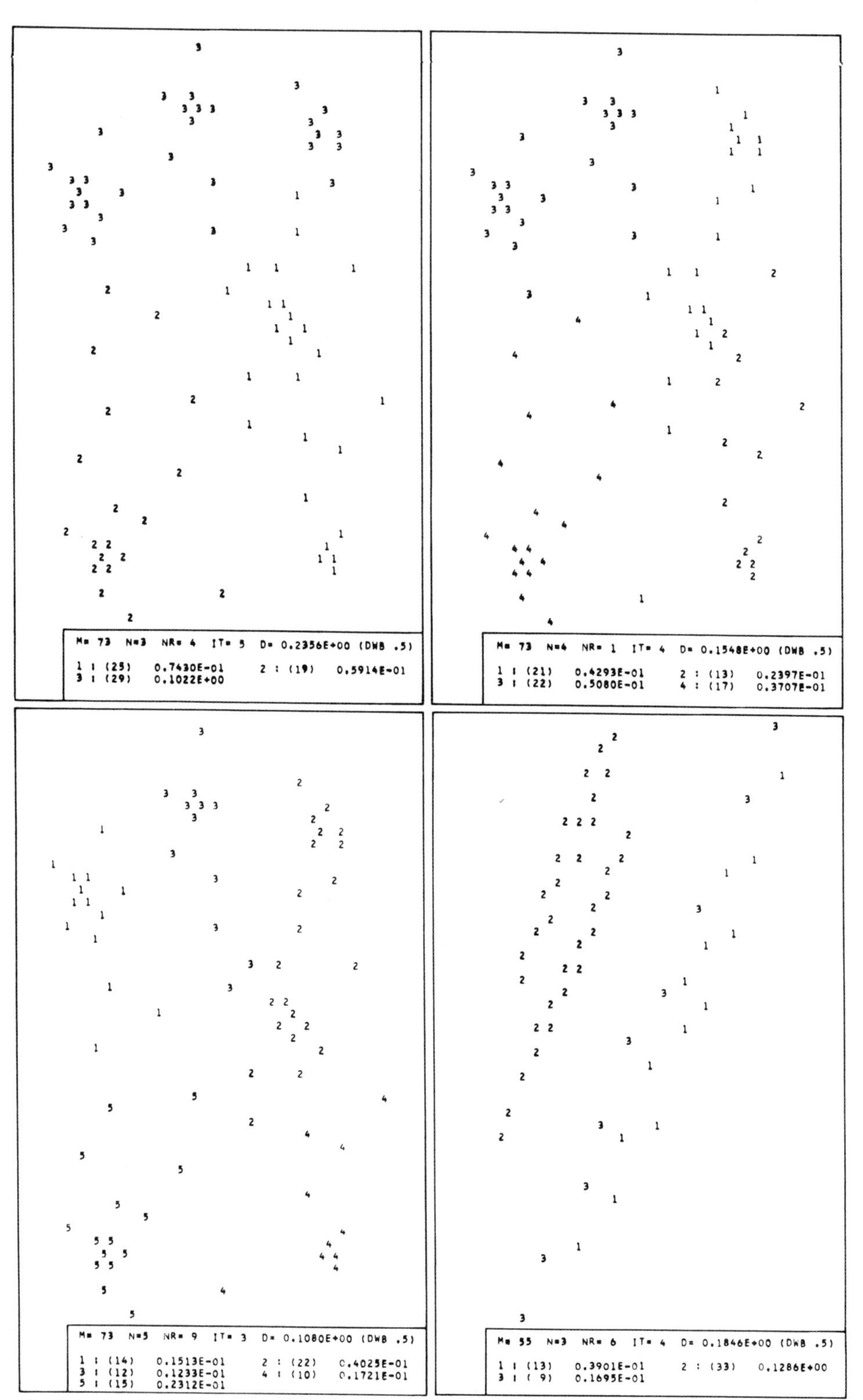
M= 73 N=3 NR= 4 IT= 5 D= 0.2356E+00 (DWB .5)
1 : (25) 0.7430E-01 2 : (19) 0.5914E-01
3 : (29) 0.1022E+00
M= 73 N=4 NR= 1 IT= 4 D= 0.1548E+00 (DWB .5)
1 : (21) 0.4293E-01 2 : (13) 0.2397E-01
3 : (22) 0.5080E-01 4 : (17) 0.3707E-01
M= 73 N=5 NR= 9 IT= 3 D= 0.1080E+00 (DWB .5)
1 : (14) 0.1513E-01 2 : (22) 0.4025E-01
3 : (12) 0.1233E-01 4 : (10) 0.1721E-01
5 : (15) 0.2312E-01
M= 55 N=3 NR= 6 IT= 4 D= 0.1846E+00 (DWB .5)
1 : (13) 0.3901E-01 2 : (33) 0.1286E+00
3 : ( 9) 0.1695E-01

*Fig. 11.12*

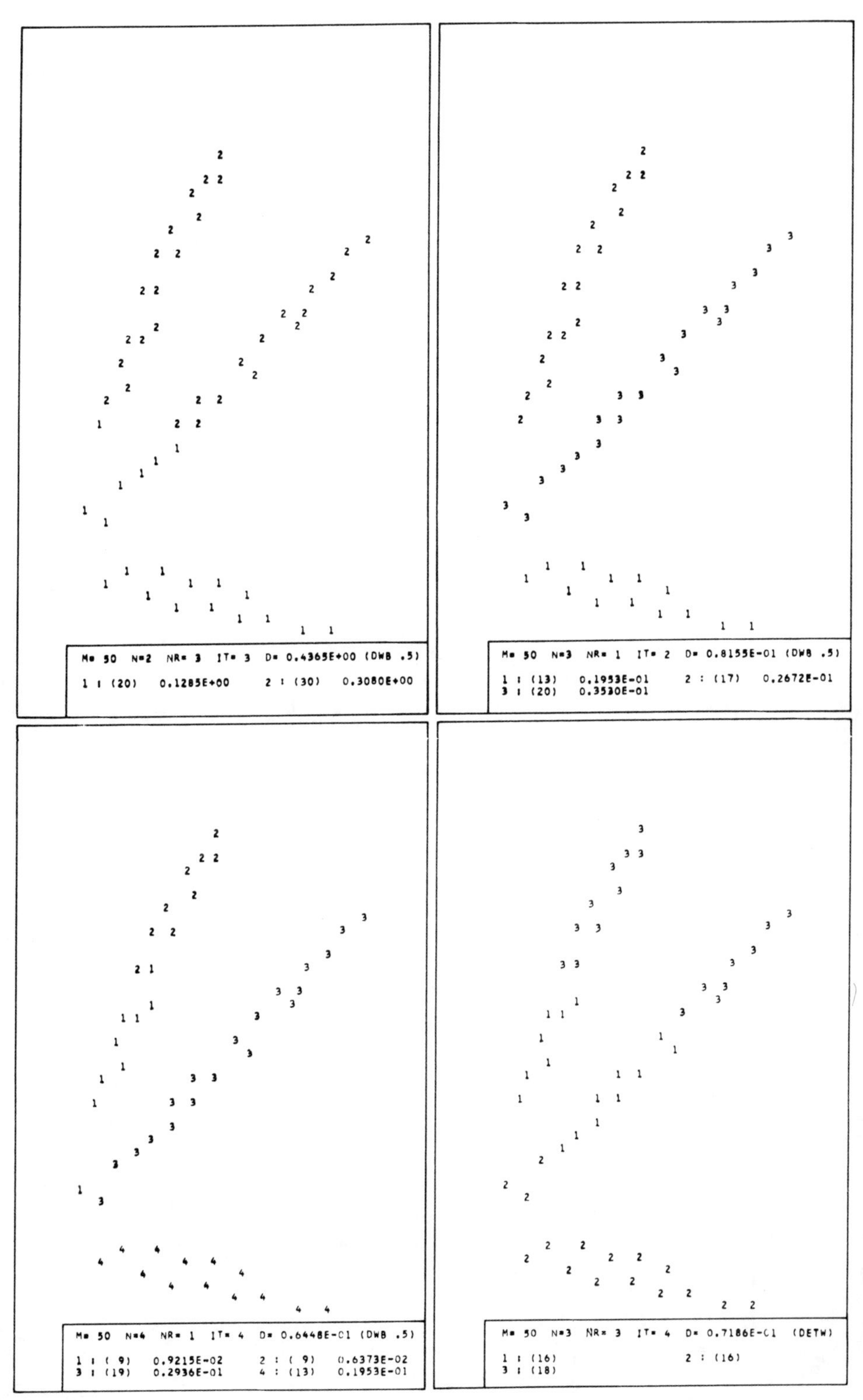
M= 50 N=2 NR= 3 IT= 3 D= 0.4365E+00 (DWB .5)
1 : (20) 0.1285E+00 2 : (30) 0.3080E+00
M= 50 N=3 NR= 1 IT= 2 D= 0.8155E-01 (DWB .5)
1 : (13) 0.1953E-01 2 : (17) 0.2672E-01
3 : (20) 0.3530E-01
M= 50 N=4 NR= 1 IT= 4 D= 0.6448E-01 (DWB .5)
1 : ( 9) 0.9215E-02 2 : ( 9) 0.6373E-02
3 : (19) 0.2936E-01 4 : (13) 0.1953E-01
M= 50 N=3 NR= 3 IT= 4 D= 0.7186E-01 (DETW)
1 : (16) 2 : (16)
3 : (18)

*Fig. 11.13*

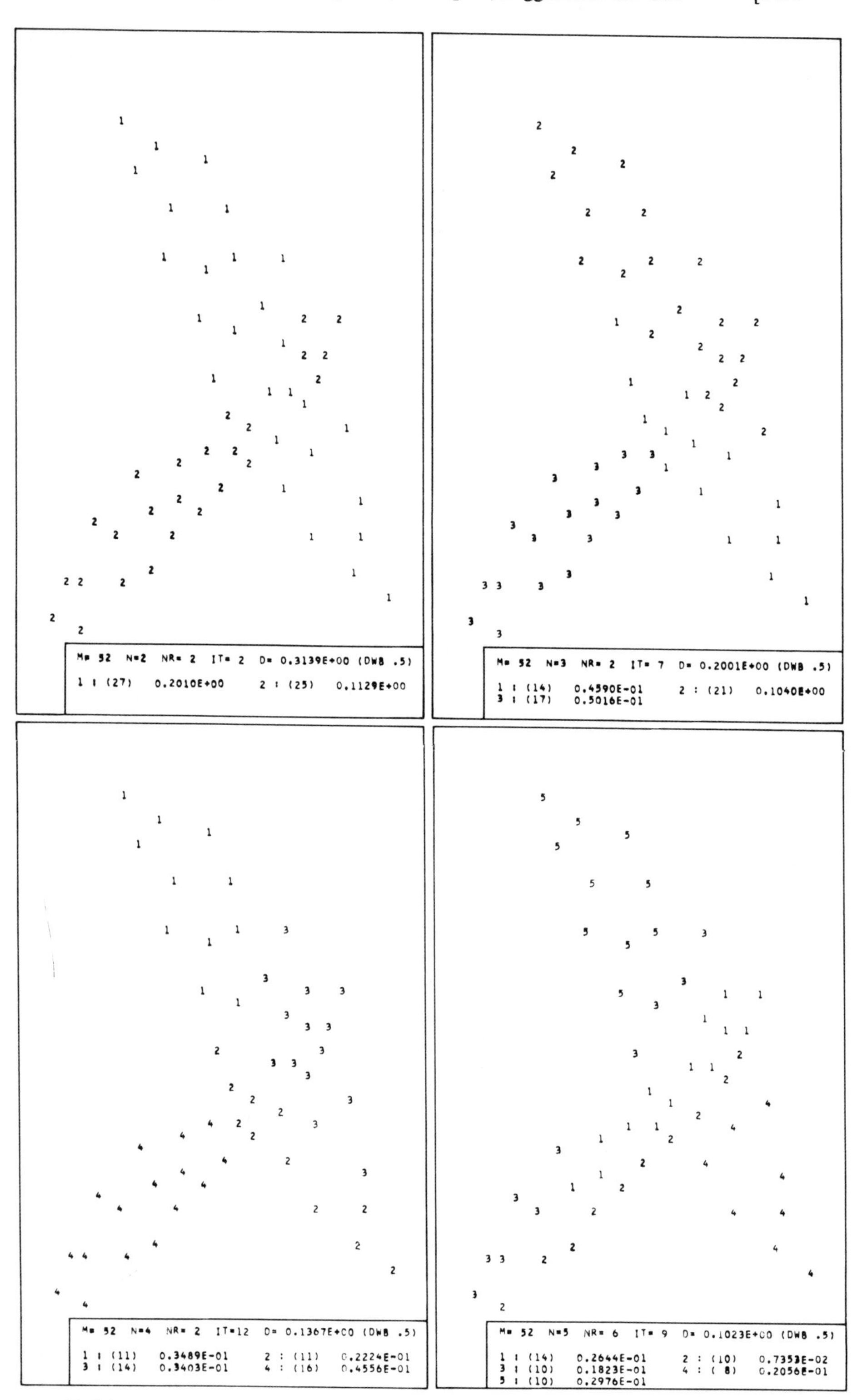
M= 52 N=2 NR= 2 IT= 2 D= 0.3139E+00 (DWB .5)
1 : (27) 0.2010E+00 2 : (25) 0.1129E+00
M= 52 N=3 NR= 2 IT= 7 D= 0.2001E+00 (DWB .5)
1 : (14) 0.4590E-01 2 : (21) 0.1040E+00
3 : (17) 0.5016E-01
M= 52 N=4 NR= 2 IT=12 D= 0.1367E+00 (DWB .5)
1 : (11) 0.3489E-01 2 : (11) 0.2224E-01
3 : (14) 0.3403E-01 4 : (16) 0.4556E-01
M= 52 N=5 NR= 6 IT= 9 D= 0.1023E+00 (DWB .5)
1 : (14) 0.2644E-01 2 : (10) 0.7353E-02
3 : (10) 0.1823E-01 4 : ( 8) 0.2056E-01
5 : (10) 0.2976E-01

*Fig. 11.14*

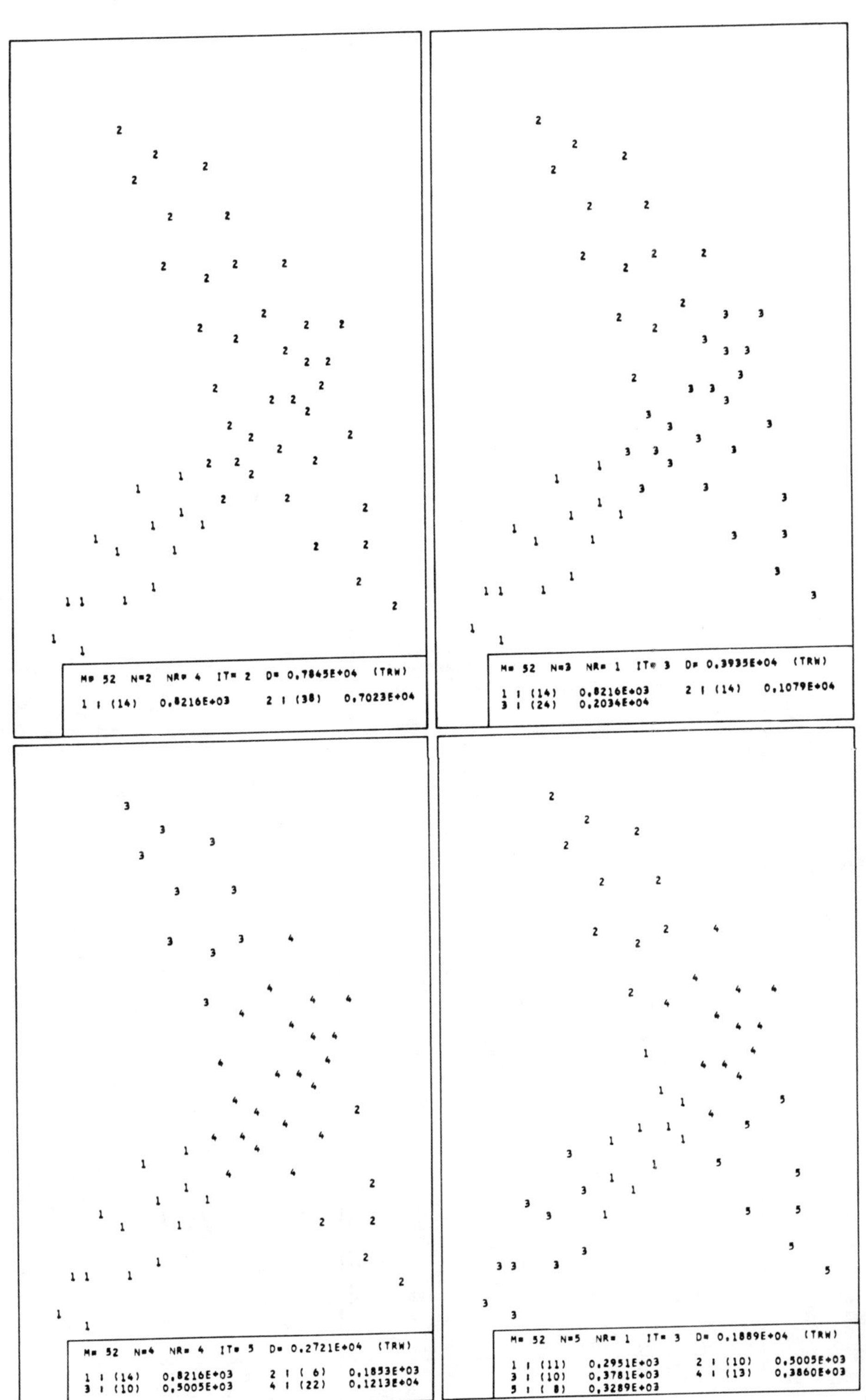
M= 52 N=2 NR= 4 IT= 2 D= 0,7845E+04 (TRW)
1 I (14) 0,8216E+03 2 I (38) 0,7023E+04
M= 52 N=3 NR= 1 IT= 3 D= 0,3935E+04 (TRW)
1 I (14) 0,8216E+03 2 I (14) 0,1079E+04
3 I (24) 0,2034E+04
M= 52 N=4 NR= 4 IT= 5 D= 0,2721E+04 (TRW)
1 I (14) 0,8216E+03 2 I ( 6) 0,1853E+03
3 I (10) 0,5005E+03 4 I (22) 0,1213E+04
M= 52 N=5 NR= 1 IT= 3 D= 0,1889E+04 (TRW)
1 I (11) 0,2951E+03 2 I (10) 0,5005E+03
3 I (10) 0,3781E+03 4 I (13) 0,3860E+03
5 I ( 8) 0,3289E+03

*Fig. 11.15*

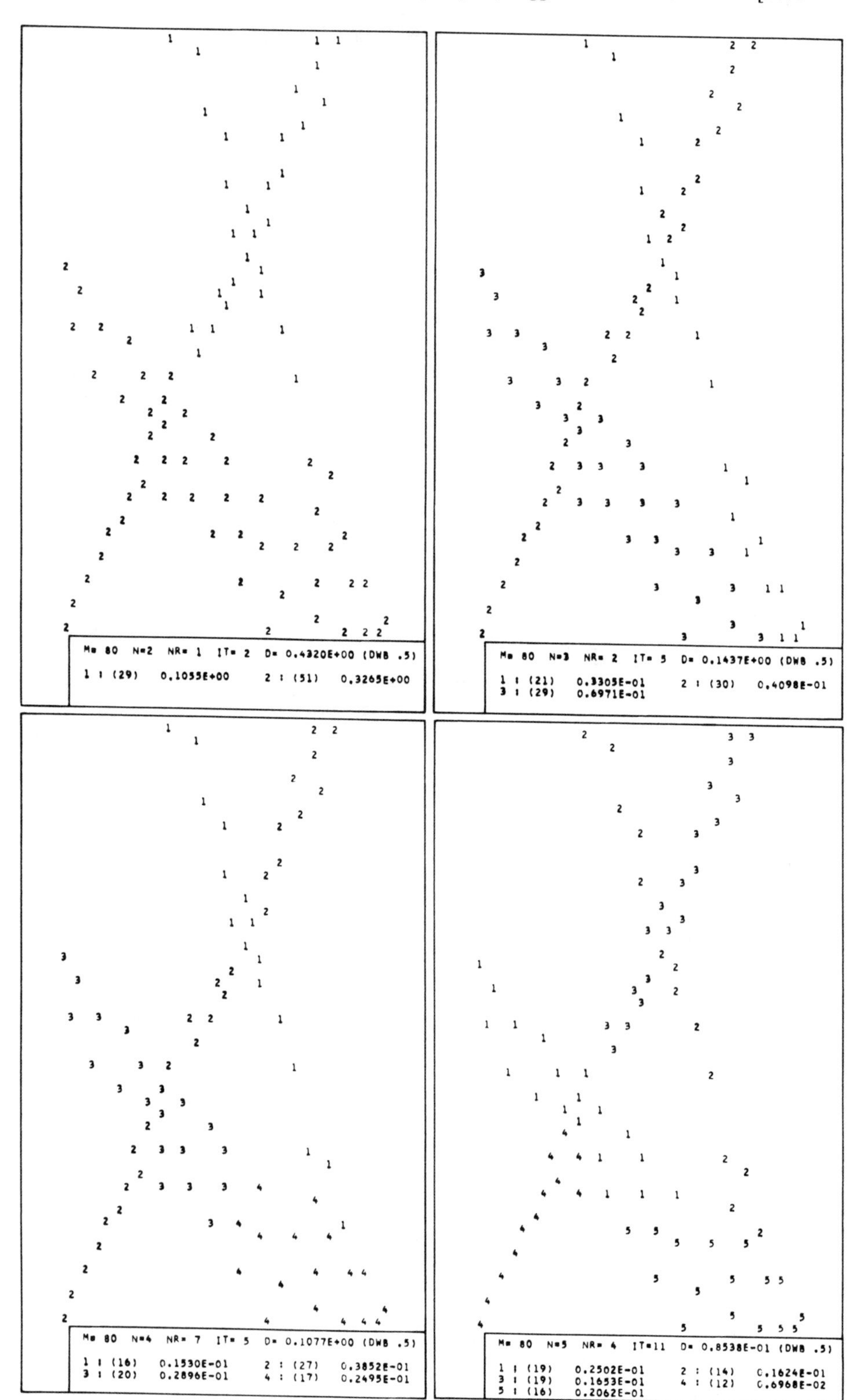
M= 80 N=2 NR= 1 IT= 2 D= 0.4320E+00 (DWB .5)
1 : (29) 0.1055E+00 2 : (51) 0.3265E+00
M= 80 N=3 NR= 2 IT= 5 D= 0.1437E+00 (DWB .5)
1 : (21) 0.3305E-01 2 : (30) 0.4098E-01
3 : (29) 0.6971E-01
M= 80 N=4 NR= 7 IT= 5 D= 0.1077E+00 (DWB .5)
1 : (16) 0.1530E-01 2 : (27) 0.3852E-01
3 : (20) 0.2896E-01 4 : (17) 0.2495E-01
M= 80 N=5 NR= 4 IT=11 D= 0.8538E-01 (DWB .5)
1 : (19) 0.2502E-01 2 : (14) 0.1624E-01
3 : (19) 0.1653E-01 4 : (12) 0.6968E-02
5 : (16) 0.2062E-01

*Fig. 11.16*

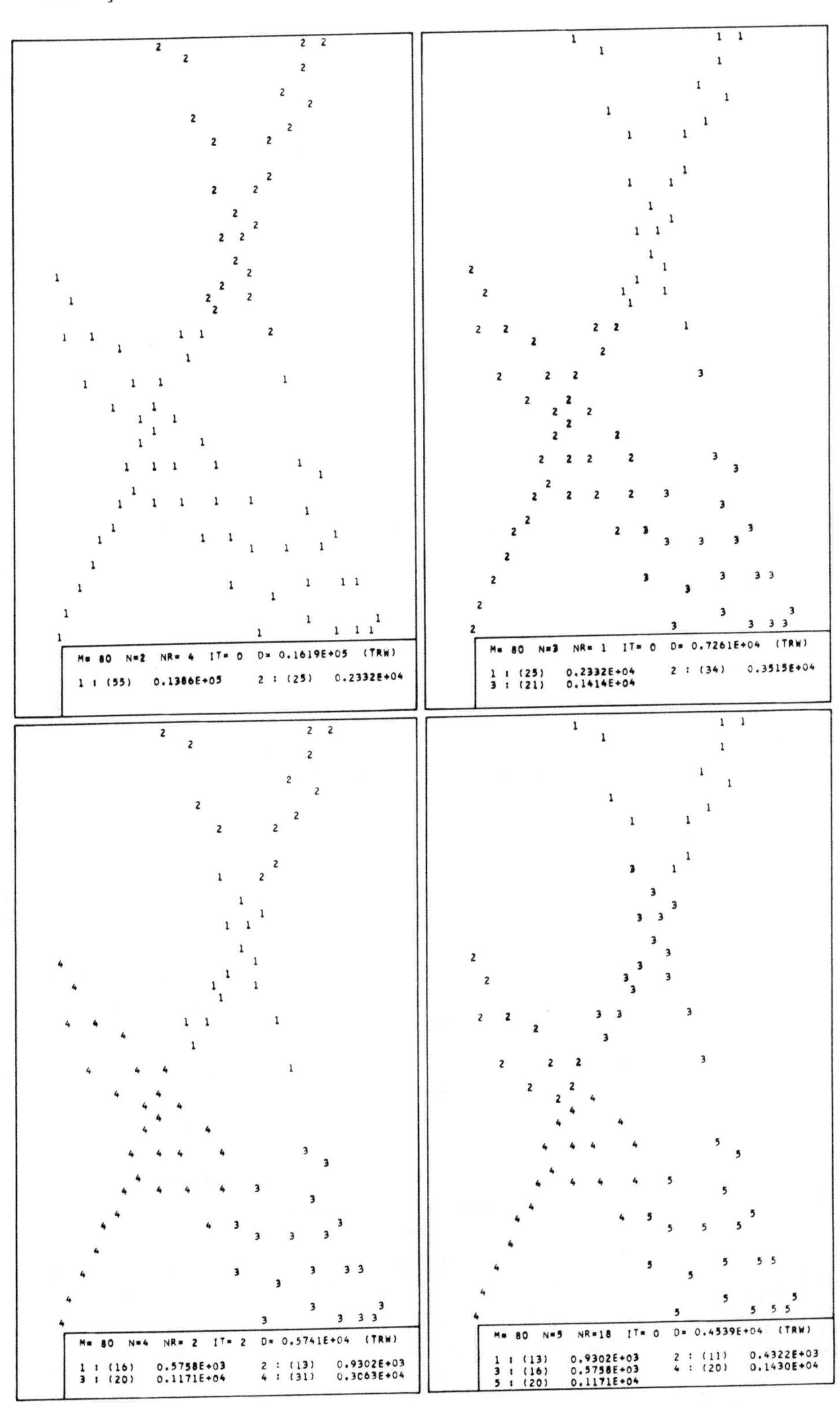
M= 80 N=2 NR= 4 IT= 0 D= 0.1619E+05 (TRW)
1 : (55) 0.1386E+05 2 : (25) 0.2332E+04
M= 80 N=3 NR= 1 IT= 0 D= 0.7261E+04 (TRW)
1 : (25) 0.2332E+04 2 : (34) 0.3515E+04
3 : (21) 0.1414E+04
M= 80 N=4 NR= 2 IT= 2 D= 0.5741E+04 (TRW)
1 : (16) 0.5758E+03 2 : (13) 0.9302E+03
3 : (20) 0.1171E+04 4 : (31) 0.3063E+04
M= 80 N=5 NR=18 IT= 0 D= 0.4539E+04 (TRW)
1 : (13) 0.9302E+03 2 : (11) 0.4322E+03
3 : (16) 0.5758E+03 4 : (20) 0.1430E+04
5 : (20) 0.1171E+04

*Fig. 11.17*

Fig. 11.13 contains results for example 6, using DWBEXM and n = 2, 3, 4 classes, as well as one result using DETEXM with n = 3. The latter differs little from that produced by TRWEXM.

For examples 7 and 8, the results using TRWEXM and DWBEXM are juxtaposed in Figs. 11.14, 11.15, 11.16, and 11.17. In example 7 with n = 2 and example 8 with n = 3, 4, 5, DWBEXM finds clusters which penetrate one another. Unlike TRWEXM, the convex hulls of the cluster members are not disjoint. Instead, they overlap. On the other hand, it can be seen that, where the number of classes differs from the 'natural' number, the criterion of adaptive distances does tend to produce clusters which can be separated by hyperplanes, even though these may differ from those associated with TRWEXM. With unstructured data, DWBEXM has a tendency to avoid clusters that penetrate each other.

The computer time required for the twelve examples already mentioned using DWBEXM was 1504 seconds, compared with 1522 seconds when using DETEXM under the same conditions. At first, it seems surprising that DWBEXM should not have used more time than DETEXM. For although each exchange attempt uses $O(mns^2)$ operations (as with DETEXM), a lot of data is moved in memory. The reason for the faster execution of DWBEXM is likely to be the greater number of attempts that were broken off prematurely owing to errors being flagged.

## 11.9 Cluster Analysis versus Cluster Dissection; Choice of Method and of the Number of Classes

In the following, we shall assume that the exchange method will often find an optimal partition, provided that sufficiently many random starting partitions are tried. It is plausible that this tends to be the case when a smallest value of the objective function occurs more frequently than other values. In general, this has been the case with our eight examples, with NRMAX = 20 starting partitions, and with all criteria. Of course we can draw no generally valid conclusions from only eight examples. However, we can formulate a number of suggestions which are in accord with the theory as well as with the author's experience in practical applications over many years.

From the way it is conceived, the variance criterion tends to find or form clusters whose shape is spherical, though their content depends on the scales used to measure the variables. The determinant criterion (which is invariant with respect to changes in scale) tends to form ellipsoids with equal axis orientations. The criterion of adaptive distances tends to form ellipsoids whose axes point in different directions.

It is therefore not surprising that the existing structures were correctly found by all three methods in the case of example 1 (spheres are special cases of ellipsoids), by the determinant criterion and the criterion of adaptive distances in the case of example 5, and by the criterion of adaptive distances in the case of examples 6 to 8. The structures were found where their shape was consistent with the concept underlying the method used, and where the number of classes was chosen correctly. (Note that, for the variance criterion, the 'correct' number of classes may depend on the scale used, as was shown in Fig. 2.1).

If one suspects or wants to admit clusters of a certain geometric shape, one must choose one of the three objective functions (or some other) which corresponds to the shape. If one does not know the 'natural' number of classes, one can vary the number of classes within certain limits, and note the best values of the objective function (and their frequency) when using several random starting partitions for the exchange method.

This we have done for our eight examples and three criteria. In Table 11.16 the resulting best values of the objective functions for $n = 2, \ldots, 6$ classes are expressed as percentages of the corresponding values for $n = 1$ class. For $n = 1$, DWBEXM returns the value $(\det W)^{1/s}$, and DETEXM the value $\det W$. To allow a direct comparison, the square root ($s = 2$) of values of the objective function was taken for DETEXM. We point out once more that the values for TRWEXM in such a table depend on the scale of measurement used (see Fig. 2.1), which is not the case with the other criteria.

With the first example, the great drop in the values from $n = 2$ to $n = 3$ using all three criteria allows us to recognise the 'right' number of classes as $n = 3$. Similarly, in example 5, the fall in DETEXM and DWBEXM from $n = 1$ to $n = 2$ and from $n = 2$ to $n = 3$ is very large, so that one could argue for $n = 2$ or $n = 3$ classes. This is consistent with Fig. 11.10. The same applies with example 6 and DWBEXM. For example 7 and DWBEXM it is clear that $n = 2$ is the 'right' number of classes. In example 4, the last sizeable drop in values for all criteria can be seen in the transition from $n = 3$ to $n = 4$, so that one should plead for $n = 4$. In example 2, which is devoid of a structure, the values decrease gradually without major jumps. The same happens with example 3, though at a lower level.

Thus, it is possible to infer the existence of a structure and the appropriate number of classes when large drops occur in the values such as those in Table 11.16. In that case, the term 'cluster analysis' is justified, where the reduction in values is gradual, we should be content with terms such as 'cluster dissection' or 'appropriate division into classes'. In this case, the number of classes is often determined by considerations of usefulness, or in an arbitrary fashion. Here, the different methods impose a structure corresponding to their mathematical properties. We stress again that sufficiently many starting partitions should be used before compiling a table such as 11.16 and drawing conclusions from it. If this is not done, some of the end partitions being compared might be less good than others, and the resulting (wrong) jumps could lead to wrong conclusions regarding the number of classes.

Thus, if one suspects a structure, one should try the three criteria, and perhaps others as well, for different numbers of classes. If one is merely interested in a subdivision into classes, one will probably use the variance criterion, possibly with several scalings, or with the standard scaling provided by TRAFOR. Where scaling presents a serious problem, it is worth considering the determinant criterion, which will also produce clusters such that the convex hulls of the cluster members are disjoint. Although DWBEXM did produce disjoint clusters in the corresponding examples (that is, those excluding numbers 7 and 8), this cannot be guaranteed. This and the high computation time make DWBEXM unsuitable for cluster dissection.

*Table 11.16*

| Cluster number | 2 | 3 | 4 | 5 | 6 |
|---|---|---|---|---|---|
| Example | | | TRWEXM | | |
| 1 | 46.8 | 13.9 | 8.6 | 6.7 | 4.5 |
| 2 | 60.0 | 38.7 | 24.9 | 19.7 | 16.7 |
| 3 | 41.4 | 24.6 | 18.6 | 14.5 | 11.9 |
| 4 | 56.9 | 31.0 | 19.4 | 13.8 | 11.6 |
| 5 | 55.7 | 31.5 | 23.0 | 15.6 | 12.2 |
| 6 | 54.0 | 33.8 | 21.3 | 14.9 | 11.2 |
| 7 | 56.5 | 28.3 | 19.6 | 13.6 | 10.6 |
| 8 | 57.3 | 25.6 | 20.9 | 16.4 | 12.4 |
| | | | DETEXM | | |
| 1 | 37.0 | 14.0 | 7.5 | 6.0 | 4.5 |
| 2 | 48.9 | 32.3 | 23.7 | 19.2 | 16.1 |
| 3 | 41.9 | 27.2 | 19.3 | 14.1 | 12.1 |
| 4 | 42.5 | 27.4 | 17.7 | 13.3 | 10.5 |
| 5 | 31.9 | 16.9 | 11.0 | 7.7 | 6.4 |
| 6 | 48.0 | 27.1 | 18.4 | 12.9 | 8.9 |
| 7 | 47.2 | 27.7 | 19.0 | 13.7 | 10.5 |
| 8 | 50.1 | 25.9 | 20.6 | 15.2 | 12.1 |
| | | | DWBEXM | | |
| 1 | 23.3 | 13.0 | 6.7 | 5.1 | - |
| 2 | 48.4 | 32.6 | 22.2 | 17.7 | 13.6 |
| 3 | 40.3 | 24.9 | 15.8 | 12.8 | 9.0 |
| 4 | 41.1 | 23.6 | 15.5 | 10.8 | 8.0 |
| 5 | 26.8 | 19.7 | 9.8 | 7.1 | 6.3 |
| 6 | 44.1 | 8.2 | 6.5 | 5.2 | 5.2 |
| 7 | 31.6 | 20.1 | 13.7 | 10.3 | 9.1 |
| 8 | 43.2 | 14.4 | 10.8 | 8.5 | 6.6 |

**Problem 11.1:** Using URAND or some other random number generator, write a subroutine for generating random starting partitions which does not assign each object randomly to a class, but instead assigns the class numbers successively to random objects. Is there any point in doing this?

**Problem 11.2:** Use the four additional examples on the magnetic tape described in Chapter 13 to compare TRWEXM, DETEXM and DWBEXM, and make up additional examples of your own. To evaluate the results, implement a subroutine to generate tables such as Table 11.16.

**Problem 11.3**: Study the Branch and Bound method for the variance criterion [A8, B26, B28] and use the subroutine described in [B26] for comparison with TRWEXM. Note down computation times and compare the value of the objective function for end partitions achieved with TRWEXM with the corresponding value for an optimal partition (which is what the Branch and Bound method in fact generates). What conclusions do you draw?

**Problem 11.4**: The numerical results obtained with TRWEXM, DETEXM and DWBEXM for larger numbers of classes suggest that NRMAX should be increased with the number of classes n if better sub-optima are to be obtained. If this turns out to require too much computer time where there is a series of large numbers of classes, then an alternative approach is possible. Starting from a good end partition for a lower number of classes $n_0 \geqslant 1$, one can use the following as starting partitions for the subsequent numbers of classes $n_h = n_{0+h}$ $(h = 1, \ldots, h')$, and then apply the exchange method, for example with TRWEXM:

$$C'' = (C_1, \ldots, C_p - \{i\}, \ldots, C_{n_h - 1}, C_{n_h} := \{i\})$$

with

$$e(C_p) = \max_{j=1,\ldots,n_h-1} e(C_j)$$

and

$$\| x_i - \bar{x}_p \| = \max_{k \epsilon C_p} \| x_k - \bar{x}_p \|$$

In analogy to the proof of monotonicity of the exchange method, this choice of starting partition guarantees a reduction in the values of the objective function as n increase. Implement this method for the variance criterion, gather some numerical experience, and consider how you would need to proceed with the determinant criterion and the criterion of adaptive distances.

**Problem 11.5**: You have a data matrix which you cannot fully process according to the criterion you have chosen, either because of its size or because of constraints on computer time. What do you do?

**Problem 11.6**: With a slight modification of the main programs for TRWEXM, DETEXM and DWBEXM, the relevant values of the objective function DS and DE for the random starting and for the end partition can be printed. Verify the empirical observation that larger or smaller values of the starting partitions are not correlated with larger or smaller values of the end partitions, for example by constructing scatter diagrams of the pairs of values (DS, DE). Consequently, there is no point in selecting from a set of random starting partitions those with the smallest value DS and performing the exchange method only for these partitions.

# 12 Sample main program and evaluation for OVSEXM, OVREXM, OVPEXM, BVPEXM, TIHEXM, and CLREXM

## 12.1 Sample Main Program and a Test Example for the $L_1$-criterion: Comparison of the three versions OVSEXM, OVREXM and OVPEXM

It would have been possible to employ the previously used examples as test examples for the $L_1$-criterion as well. Since their data matrices consisted of integers, it would even have been possible to use the subroutines OVSEXM and OVREXM exactly in the form in which they have been listed, without the need to convert them for use with real data matrices, as described in Section 10.1. However, we shall use a different example since we intend to make comparisons with OVPEXM where an integer data matrix and a smaller number T of different integers are necessary conditions (the former in theory, the latter in practice). The example has a data matrix X of m = 24 objects and s = 122 variables, which is shown in transposed form in Table 12.1. The m objects were items on a psychological questionnaire [B7] which were originally filled in by 260 people using an ordinal scale with values ranging from 1 to T = 4. To save space, we have restricted ourselves to the first 122 of these persons. Later on, when we refer to a case with s = 60, we shall mean the first 60 of these 122 persons. Further examples for comparison of computing times, but also of practical scientific interest, were obtained by exchanging the roles of m and s, that is, by forming clusters of people rather than clusters of questions. In this case, the variables were the values of the answers for the different questions.

The main program of OVSEXM shown in Table 12.2 is set up for a maximum of M = 24, S = 260 and N = 6. The data matrix in Table 12.1 is read in its transposed form (S = 122), and the computation is performed for NRMAX = 20 random starting partitions and N = N1 = N2 = 4 different classes.

As previously, the output is divided into two parts. Each row of Table 12.3 shows, for one of the NR = 1, NRMAX starting partitions, the values of M, N, NR, IFLAG, IT, D, (J, MJ(J), E(J), J = 1, N). The exchange method shows the usual behaviour regarding the values of the objective function D. In contrast to previous chapters, Table 12.4 lists only the best end partition, or the first of these if there are several (as here, with NR = 3), and identifies the partition by the serial numbers of the class members, and by the values of the variables for the class members.

The main programs for the other versions OVREXM and OVPEXM are nearly identical to that for OVSEXM. Table 12.5 shows the differences. If the main programs are edited according to the table, they will produce the same output with the given example as the program using OVSEXM.

*Table 12.1*

| | |
|---|---|
| 1 | 143111122123343413311111 |
| 2 | 141311121113422112121112 |
| 3 | 233412212334343433342211 |
| 4 | 231212121214311322341212 |
| 5 | 332311111114413111111111 |
| 6 | 432311121313421421211114 |
| 7 | 333412112434334343443241 |
| 8 | 334321111134444433432323 |
| 9 | 142412311324343412341111 |
| 10 | 443313211134444344433331 |
| 11 | 443422112124443332312211 |
| 12 | 443421123344424444443442 |
| 13 | 313143112134334432331132 |
| 14 | 313241141221232143243444 |
| 15 | 344341141224413211322144 |
| 16 | 342412321114342413121421 |
| 17 | 131443441411111124341143 |
| 18 | 313143112134334432321212 |
| 19 | 131211311124313122141432 |
| 20 | 344321111114433331332231 |
| 21 | 343421112224444432432222 |
| 22 | 334211111123422332311111 |
| 23 | 343421232124423133441343 |
| 24 | 443212212224433333333334 |
| 25 | 343211111223433332222322 |
| 26 | 344322211123433332322111 |
| 27 | 443323211124433423322111 |
| 28 | 434421121224433332211121 |
| 29 | 343221123124442311223111 |
| 30 | 443312212224442322222222 |
| 31 | 344321111124433331331111 |
| 32 | 342331111114432321121121 |
| 33 | 444421212223433332332122 |
| 34 | 342111211124432231311121 |
| 35 | 333321212223433332222221 |
| 36 | 333111221223432221112222 |
| 37 | 444321111113432332321131 |
| 38 | 332321111114321121332112 |
| 39 | 333212111214333332332111 |
| 40 | 333212111223333322221111 |
| 41 | 443311112124433232312111 |
| 42 | 433322121223433232322222 |
| 43 | 433321121124332221422111 |
| 44 | 433321121224432221231111 |
| 45 | 343321123224433332321111 |
| 46 | 444321111214443331311132 |
| 47 | 212343444441212134334344 |
| 48 | 222342344341133344444244 |
| 49 | 444443344434443343443343 |
| 50 | 333241244133422211232342 |
| 51 | 333321111224433332321221 |
| 52 | 332331234124423232231232 |
| 53 | 444311113133421131441131 |
| 54 | 333331244123311111111133 |
| 55 | 333331244223432322231133 |
| 56 | 323341143124432221331132 |
| 57 | 443311133124432331232121 |
| 58 | 324431443232432312243343 |
| 59 | 443341134134422221332131 |
| 60 | 333241233133412131212142 |
| 61 | 231343443421312122343143 |
| 62 | 323241133323422322232222 |
| 63 | 332221132213331232322222 |
| 64 | 333342342113322231331142 |
| 65 | 343311211124433432312111 |
| 66 | 332211111124324321311121 |
| 67 | 434312324114344344341224 |
| 68 | 343412211124434332221121 |
| 69 | 143421121434344321341212 |
| 70 | 443311111114443111211131 |
| 71 | 421421112222324222222111 |
| 72 | 332311112114432112212111 |
| 73 | 143421121134434131423112 |
| 74 | 433411212124323321332112 |
| 75 | 434311111234243432421321 |
| 76 | 332211111124324321311111 |
| 77 | 343241221422313111121142 |
| 78 | 443311111224433421321121 |
| 79 | 444432212324444443434322 |
| 80 | 443422221124422422342211 |
| 81 | 343411212134342321231212 |
| 82 | 332311112113322112212121 |
| 83 | 343441313323433443432343 |
| 84 | 434421121114433412312111 |
| 85 | 343341231324433111342243 |
| 86 | 343342233224433332322121 |
| 87 | 433321213224433432332223 |
| 88 | 444312221123333432322121 |
| 89 | 444321111134433433322112 |
| 90 | 444422211234433332211211 |
| 91 | 344331343314422111231143 |
| 92 | 434321111213433332221121 |
| 93 | 443321112124433331231121 |
| 94 | 443312312423444424442214 |
| 95 | 443321223424433432431221 |
| 96 | 443322211223433332421112 |
| 97 | 444433323334442443443322 |
| 98 | 434432323324433332432122 |
| 99 | 344321121124442332421111 |
| 100 | 344322211214433332321232 |
| 101 | 343442222224432323322122 |
| 102 | 343321111123322232312121 |
| 103 | 443322121214433442332211 |
| 104 | 443341132224443432322111 |
| 105 | 444412211134433432321111 |
| 106 | 343322212234422222222121 |
| 107 | 443433122224213421212121 |
| 108 | 343341123223432332231222 |
| 109 | 433211233324432312342132 |
| 110 | 343341143224423121331142 |
| 111 | 443321233234433342332223 |
| 112 | 433322211124443333322111 |
| 113 | 433321121114422121222122 |
| 114 | 333313111224333332332232 |
| 115 | 434332133224442332432132 |
| 116 | 333311121114433241431111 |
| 117 | 443321222214443344221211 |
| 118 | 344321121113433331311111 |
| 119 | 443311111124443332221111 |
| 120 | 443331112224443333421121 |
| 121 | 344331132124422221221212 |
| 122 | 444442244444444442442444 |

*Table 12.2*

```
C   SAMPLE MAIN PROGRAM FOR OVSEXM
C
C
       INTEGER X(24,260),U(6,260),V(6,260)
       INTEGER XX(24,260),MK(7),SP(260),SJ(260),SQ(260)
       INTEGER S,D,MJ(6),E(6),Z(24),P(24)
       INTEGER UP(260),VP(260),UJ(260),VJ(260),UQ(260),VQ(260),Y(24)
       LOGICAL EVEN(6)
C
       N1=4
       N2=4
       KBIG=99999999
       M=24
       S=122
       NRMAX=20
       KI=5
       KO=6
       WRITE(KO,20)
       DO 2 K=1,S
            READ(KI,3)    (X(I,K),I=1,M)
            WRITE(KO,4)K,(X(I,K),I=1,M)
     2 CONTINUE
     3 FORMAT(3X,24I1)
     4 FORMAT(' ',I4,3X,24I1)
       DO 9 N=N1,N2
            NN=N+1
            IY=25712369
            WRITE(KO,20)
            KD=KBIG
            DO 6 NR=1,NRMAX
                 CALL RANDP (M,N,Z,IY)
                 CALL OVSEXM (X,24,M,S,Z,MJ,6,N,U,V,E,D,
     *                        IT,IFLAG,Y,EVEN,UP,VP,UJ,VJ,UQ,VQ,
     *                        SP,SJ,SQ,NN,MK,XX)
                 IF(IFLAG.NE.0) GOTO 6
                 WRITE(KO,22) M,N,NR,IFLAG,IT,D,
     *                        (J,MJ(J),E(J),J=1,N)
                 IF(D.GE.KD) GOTO 6
                 KD=D
                 DO 5 I=1,M
                      P(I)=Z(I)
     5           CONTINUE
     6      CONTINUE
            WRITE(KO,23) N,NRMAX
            WRITE(KO,25)
            DO 8 J=1,N
                 WRITE(KO,24) J
                 WRITE(KO,25)
                 DO 7 I=1,M
                      IF(J.NE.P(I)) GOTO 7
                      WRITE(KO,21) I,(X(I,K),K=1,S)
     7           CONTINUE
                 WRITE(KO,25)
     8      CONTINUE
     9 CONTINUE
    20 FORMAT('1')
    21 FORMAT('0',I3,3X,67I1/(' ',6X,67I1))
    22 FORMAT(' ',I2,1X,I1,1X,3I2,I8,3X,9(I2,I3,I7))
    23 FORMAT('1','BEST SOLUTION FOR',I2,' CLUSTERS AMONG ',I2,
     *        ' TRIALS')
    24 FORMAT(' ','CLUSTER ',I1)
    25 FORMAT('0')
       STOP
       END
```

*Table 12.3*

```
24 4  1 0 2   1238   1  8   427 2  4   148 3  5   288 4  7   375
24 4  2 0 2   1316   1 13   862 2  5   262 3  3    80 4  3   112
24 4  3 0 3   1234   1  7   351 2  7   409 3  5   214 4  5   260
24 4  4 0 3   1234   1  7   409 2  5   214 3  5   260 4  7   351
24 4  5 0 3   1234   1  5   260 2  7   351 3  5   214 4  7   409
24 4  6 0 3   1288   1 10   621 2  7   375 3  4   148 4  3   144
24 4  7 0 1   1399   1  6   286 2  3   181 3 10   650 4  5   282
24 4  8 0 3   1236   1  5   288 2  8   425 3  4   148 4  7   375
24 4  9 0 4   1238   1  7   375 2  4   148 3  8   427 4  5   288
24 4 10 0 1   1395   1  3   151 2 11   734 3  7   390 4  3   120
24 4 11 0 4   1234   1  7   409 2  5   214 3  5   260 4  7   351
24 4 12 0 2   1325   1 12   764 2  5   285 3  3   128 4  4   148
24 4 13 0 2   1408   1  3   112 2  8   475 3 10   650 4  3   171
24 4 14 0 4   1238   1  4   148 2  7   375 3  8   427 4  5   288
24 4 15 0 4   1238   1  4   148 2  5   288 3  8   427 4  7   375
24 4 16 0 4   1234   1  7   351 2  5   214 3  7   409 4  5   260
24 4 17 0 2   1350   1  9   559 2  3   123 3  7   359 4  5   309
24 4 18 0 2   1373   1  5   281 2  3   123 3  5   235 4 11   734
24 4 19 0 4   1238   1  8   427 2  4   148 3  7   375 4  5   288
24 4 20 0 1   1234   1  7   351 2  7   409 3  5   260 4  5   214
```

*Table 12.4*

```
BEST SOLUTION FOR 4 CLUSTERS AMONG 20 TRIALS

CLUSTER 1


   6   112211212321311233111112123112111111112212111132311111111111113112112
       21111111111221111212121112123212212122311112132111112

   7   112111113211111341311122122112112222111111111114332121221141241132I3
       211111211121222131222212311322331221112211212211112111112

  10   113213413113122141112112211212112122112212122243412111211211432II11
       1412111214231113132211232144233122122122232211221211214

  11   213111332324322113212222222222221222211122222214433223222233322II221
       23121323222223121222233112222322122123322223212211112224

  18   32321133242423134221223322321211212121222211214431221121121122212I4
       2112211211132123212223212142232223222221221231221412312

  21   112111321323132111122113222132112122122122211144321111112322322I211
       1111223211114212222222221111211321122222122121222221111112

  22   112211231324141412422133311112111122111112111132332211111311122III2
       121111131113221312121121112213112112111121121121121124

CLUSTER 2


   4   134233434344123441234242233423334131332233333333423333333343232233323
       44343443223444344333334333333443343334343233333333333334

  14   4241123444423314131342233333443333332333333341342322133332112323324
       344233242134242333333332334334343323432133234234343424

  15   3231314434344232143342333333223232322133332233233233112222222212344
       44342433433422233333333233433232322333232233332323333324

  16   4143143443344124141343133343333332323133222233133232113233211322433
       3312113431444314413444313344343333244424331331332333324
```

*Table 12.4 – continued*

```
 17   1132124314343411232333333323123233323232332233344133312231232233324
      3212132321242214113333313323343332343322312432334433324

 19   3133124434343231331343432332223133213332334233344232412322323233333
      2322243431343224333333222244444433333222333323442324 24

 20   1244114343143422424331432221223231212332122321344323413334314323114
      241212321223431314232213234324322213222134332233321 2224

CLUSTER 3

  1   1122343314443333131333343344343343334333444434224333433343432333334
      3144314433444333433444434444444433334444343434443434 34434

  2   4433333334444114431344344444344444433433343334412433343324243323 3433
      4442343334444434344344443444443444444444443443333444444

  3   3131223423333342131434333434334242334233333334224332433334331323324
      3331233423343323433344444333344443333433333333343343344

 12   3344434444444144144443443344444434333443434444114344333442431333444
      4442444442444433444434443434344444344444344444444434444

 13   3433443434443243133444444444444444444433344344421444443444444433433
      4343443233444334444434444444444444434444244444434444444

CLUSTER 4

  5   1111111211224441441221211222212321212211122222444423133413444424111
      1212121114132114244212232212233224224123414222131221334

  8   2212121111121442411111311112211111211111222214444131444343343341 12
      1211121112112111233121141112122212123112234312132221134

  9   2121112111232111121121221111321121211111211131444414344333433322114
      1112212111121223113311131223133112112122333311131211224

 23   1111114213143442413321432112121222223111121113444423333243442 24122
      2131111214221124142221142212122132211122234212331111214

 24   1212141311122441322121342111121121121211121112443212133213123222114
      1211122112121213131312131141222122111111222312221111124
```

Table 12.6 shows comparative computation times for the three programs in terms of seconds on a TR 440. The first row corresponds to our example as described. The other rows show times for the shortened, extended and transposed data matrix from Table 12.1. For S > M, OVSEXM is unbeatable in terms of speed. However, for M > S it is advisable to use OVPEXM. In both cases, the computation time required for OVREXM lies in the middle, but OVREXM requires least storage when T > 2. As can be seen from the DIMENSION statements in the three subroutines, the storage required is:

| | |
|---|---|
| OVSEXM: | 2ms + 2ns + 2m + 4n + 9s |
| OVREXM: | ms + 2ns + 2m + 3n + 6s |
| OVPEXM: | ms + tns + m + 3n + t. |

*Table 12.5*

```
C   SAMPLE MAIN PROGRAM FOR OVREXM
C
C
      INTEGER X(24,260),U(6,260),V(6,260)
      INTEGER S,D,MJ(6),E(6),Z(24),P(24)
      INTEGER UP(260),VP(260),UJ(260),VJ(260),UQ(260),VQ(260),Y(24)
      LOGICAL EVEN(6)
C
      .
      .
      .
                 CALL OVREXM (X,24,M,S,Z,MJ,6,N,U,V,E,D,
     *                  IT,IFLAG,Y,EVEN,UP,VP,UJ,VJ,UQ,VQ)
      .
      .
      .
      STOP
      END



C   SAMPLE MAIN PROGRAM FOR OVPEXM
C
C
      INTEGER X(24,260),B(4),C(6),A(4,260,6)
      INTEGER S,D,T,MJ(6),E(6),Z(24),P(24)
C
      .
      .
      .
      T=4
      .
      .
      .
                 CALL OVPEXM (X,24,M,260,S,Z,MJ,6,N,E,D,IT,
     *                  A,4,T,IFLAG,B,C)
      .
      .
      .
      STOP
      END
```

*Table 12.6*

| M | S | N | NRMAX | OVSEXM | OVREXM | OVPEXM |
|---|---|---|---|---|---|---|
| 24 | 122 | 4 | 20 | 193 | 231 | 329 |
| 24 | 60 | 4,5,6 | 20 | 337 | 439 | 586 |
| 24 | 260 | 4,5,6 | 5 | 313 | 381 | 564 |
| 260 | 24 | 4 | 5 | 381 | 675 | 217 |
| 260 | 24 | 6 | 5 | 492 | 1099 | 384 |

Thus, the user can select that implementation which is most appropriate to the storage and computer time available to him. It is apparent that some suitable compromise has to be made. (Note that the storage requirement can be halved in all cases if INTEGER*2 declarations can be used.)

## 12.2 Sample Main Program and Test Example for the $L_1$-criterion Applied to Binary Data

As our test example for BVPEXM we have chosen the binary matrix shown in Table 12.7, with m = 49 and s = 56. It is taken from an application quoted in [B41]. The objects are universities, the variables subjects which are present (one) or absent (zero).

The main program for BVPEXM listed in Table 12.8 is set up for a maximum of M = 50, S = 100 and N = 9. The given data matrix is read row by row into (B(K), K=1, S) in its binary representation. For use by BVPEXM it is then converted into a matrix X with elements .TRUE. (T) and .FALSE. (F). Apart from that, the main program is nearly identical to that used with OVSEXM.

Again, the output for each number of classes between N1=2 and N2=6 is divided into two parts. First the values of M, N, NR, IFLAG, D, (J, MJ(J), E(J), J=1, N) are printed with the same meaning as before in rows corresponding to NR=1, NRMAX, then the (first) best end partition is shown fully in terms of the object numbers and object vectors in each class. The results from our example are reproduced in the four Tables 12.9, 12.10, 12.11, 12.12 for N = 3 and N = 6. In each cluster, those rows (variables) which have contributed to the cluster dissection are clearly distinguished since they consist predominantly of T- or F-elements.

Regarding the distribution of values of the objective function, the usual behaviour of the exchange method can be seen. Since the best values of the objective function for n = 2, . . ., 6 were the numbers 336, 287, 263, 247 and 234, and since these occurred 7, 16, 1, 1, and 1 times when using 20 random starting partitions, it can be surmised that n = 3 is an appropriate number of classes in this instance.

The computation time for the example of Table 12.7, that is, for M = 49, S = 56, N = 2, ..., 6 and NRMAX = 20, was 223 seconds on the TR 440, using the main program of Table 12.8. The computation speed of BVPEXM has been measured during practical application with a 1085 × 26 data matrix, using an IBM 370/168 which is about twenty times faster than the TR 440. With M = 1085, S = 26, N = 10, 20, 30 and NRMAX = 5, a total of 737 CPU seconds were required. For M = 26, S = 1085, N = 2, . . ., 6 and NRMAX = 10, 328 CPU seconds were needed in total. However, in contrast to the version of BVPEXM given here, the arrays A(MDX, S) and Z(M) were declared as INTEGER*2, and X(MDX, S) and XIK as LOGICAL*1, to save storage.

*Table 12.7*

| | |
|---|---|
| 1 | 010110001011001000010011000111100100010100000000000100001 |
| 2 | 000000000000000000000000000000000000000000000000010000001 |
| 3 | 000000000000000000000000000000000000010000000000000010000000 |
| 4 | 000001111001000000111000000001010101011100011000001100 10 |
| 5 | 01011100000110110101101111011100010001011100000000101100 |
| 6 | 000000001001000000000000000000100000101010000100000010000 |
| 7 | 0011000010010010001100000000011000000010100010001100000001 |
| 8 | 101001001001000000110101000000101010001010000000001011 0000 |
| 9 | 0101010010010010001100010000110001010111000000000000110001 |
| 10 | 00000000100100100000000000000000100000101000011000000000001 |
| 11 | 000010000000100000001101000011110010001000000000000000000000 |
| 12 | 010100001001001000010001000011101010011000000000000000100100 |
| 13 | 0101000000011010000000001000011000000101001000000100000000 01 |
| 14 | 0000000000010010000000000000011000000101000000100000000000001 |
| 15 | 000000001001000000000000000000100000101010000000000000000000 |
| 16 | 000001001001101000010001000000100010001010001000000000 10010 |
| 17 | 00010000000010000000000000000001100000101000000000000000001 |
| 18 | 000001001001000000011000000000101010101010000100000000 10010 |
| 19 | 00000000100100001011000000000010001000101000000000010010000 |
| 20 | 10000000100100000001101000000000100010001010000000000000000001 |
| 21 | 100001001001000010110000000000100110001010001000000000 10010 |
| 22 | 00000100100100001011000100000010101110111010010000000 10010 |
| 23 | 0101000110010010011100001000110101000100010000000000100001 |
| 24 | 001000000100101000000100010000001000100010100001000000010000 |
| 25 | 100000001000000000000010001000000000000000000000000000000001 |
| 26 | 0100000010010010000000001000011000000010010000000000000000000 |
| 27 | 01010000000110100001100100001101010001000000000000000110100 |
| 28 | 010100000000000000000000000011000000010000010000000000000 |
| 29 | 00000100100100000001101010010010111111010100000000000000010000 |
| 30 | 000001001001000000011000000000101010101010000000000000010010 |
| 31 | 000000001001000000000000000000100000000101000000010000000000 |
| 32 | 000000000000101000000100000000000101010101010010010111001 0010 |
| 33 | 000001000000000000000000000000100000000001000110000000 10010 |
| 34 | 000000001001000000000100000000000100010101110000100000000 10000 |
| 35 | 1101000011010011011110101110011001100010000000000000000100000 |
| 36 | 0000010010010000101110000000001010100010100001000100 10010 |
| 37 | 000001000000100000001100000000000110010101010000000000000010000 |
| 38 | 0000000010010000000000000000000100000101001001000000000000001 |
| 39 | 000000000000000000000000000000100000101000000100000000000000 |
| 40 | 00000000000100100000000000000001100000101000000000000000000001 |
| 41 | 0000010010010000000100000000001000001011100001000100 10000 |
| 42 | 0000010010010111000110001000000100010101010000110000011010 |
| 43 | 000100000000100100000000000000011000001010000000000000000001 |
| 44 | 0101010010011010001100010000011001000100000000000000100000 |
| 45 | 0000010000000000000000000000000000000100010000100000000 10000 |
| 46 | 0000000110010000000110000000000100010101010010000000 10010000 |
| 47 | 000000001001000000000000000000010000000010000000000000000000 |
| 48 | 00000100100100000001100000000000100010101010000000000 10010000 |
| 49 | 000100000000100100000000000000011000001010000001000000000001 |

*Table 12.8*

```
C   SAMPLE MAIN PROGRAM FOR BVPEXM
C
C
      INTEGER S,MJ(9),E(9),D,A(9,100),C(9),Z(50),P(50)
      INTEGER B(100)
      LOGICAL X(50,100)
C
      N1=2
      N2=6
      M=49
      S=56
      KBIG=99999999
      NRMAX=20
      IY=37519281
      KO=6
      WRITE(KO,20)
      DO 2 I=1,M
            READ(5,3) (B(K),K=1,S)
            WRITE(KO,4) I,(B(K),K=1,S)
            DO 1 K=1,S
                  X(I,K)=B(K).EQ.1
    1       CONTINUE
    2 CONTINUE
    3 FORMAT(20X,60I1)
    4 FORMAT(5X,I3,3X,60I1)
      DO 9 N=N1,N2
            WRITE(KO,20)
            KD=KBIG
            DO 6 NR=1,NRMAX
                  CALL RANDP (M,N,Z,IY)
                  CALL BVPEXM (X,50,M,S,Z,MJ,9,N,E,D,IT,IFLAG,A,C)
                  IF(IFLAG.NE.0) GOTO 6
                  WRITE(KO,22) M,N,NR,IFLAG,IT,D,
     *                         (J,MJ(J),E(J),J=1,N)
                  IF(D.GE.KD) GOTO 6
                  KD=D
                  DO 5 I=1,M
                        P(I)=Z(I)
    5             CONTINUE
    6       CONTINUE
            WRITE(KO,23) N,NRMAX
            WRITE(KO,25)
            DO 8 J=1,N
                  WRITE(KO,24) J
                  WRITE(KO,25)
                  DO 7 I=1,M
                        IF(J.NE.P(I)) GOTO 7
                        WRITE(KO,21) I,(X(I,K),K=1,S)
    7             CONTINUE
                  WRITE(KO,25)
    8       CONTINUE
    9 CONTINUE
   20 FORMAT('1')
   21 FORMAT(' ',I3,3X,100L1)
   22 FORMAT(' ',I2,1X,I1,1X,3I2,I4,3X,9(I2,I3,I4))
   23 FORMAT('1','BEST SOLUTION FOR',I2,' CLUSTERS AMONG ',I2,
     *       ' TRIALS')
   24 FORMAT(' ','CLUSTER ',I1)
   25 FORMAT('0')
      STOP
      END
```

*Table 12.9*

| | | | | | | | | | | | | | | |
|---|---|---|---|---|---|---|---|---|---|---|---|---|---|---|
| 49 | 3 | 1 | 0 | 2 | 287 | 1 | 10 | 80 | 2 | 22 | 129 | 3 | 17 | 78 |
| 49 | 3 | 2 | 0 | 2 | 287 | 1 | 10 | 80 | 2 | 19 | 106 | 3 | 20 | 101 |
| 49 | 3 | 3 | 0 | 3 | 287 | 1 | 19 | 91 | 2 | 20 | 116 | 3 | 10 | 80 |
| 49 | 3 | 4 | 0 | 2 | 287 | 1 | 20 | 116 | 2 | 10 | 80 | 3 | 19 | 91 |
| 49 | 3 | 5 | 0 | 3 | 287 | 1 | 10 | 80 | 2 | 19 | 106 | 3 | 20 | 101 |
| 49 | 3 | 6 | 0 | 2 | 287 | 1 | 10 | 80 | 2 | 20 | 114 | 3 | 19 | 93 |
| 49 | 3 | 7 | 0 | 3 | 287 | 1 | 10 | 80 | 2 | 20 | 101 | 3 | 19 | 106 |
| 49 | 3 | 8 | 0 | 5 | 304 | 1 | 21 | 124 | 2 | 15 | 114 | 3 | 13 | 66 |
| 49 | 3 | 9 | 0 | 2 | 287 | 1 | 10 | 80 | 2 | 20 | 101 | 3 | 19 | 106 |
| 49 | 3 | 10 | 0 | 2 | 287 | 1 | 10 | 80 | 2 | 20 | 101 | 3 | 19 | 106 |
| 49 | 3 | 11 | 0 | 3 | 287 | 1 | 19 | 92 | 2 | 11 | 89 | 3 | 19 | 106 |
| 49 | 3 | 12 | 0 | 2 | 287 | 1 | 20 | 101 | 2 | 19 | 106 | 3 | 10 | 80 |
| 49 | 3 | 13 | 0 | 2 | 287 | 1 | 10 | 80 | 2 | 21 | 119 | 3 | 18 | 88 |
| 49 | 3 | 14 | 0 | 3 | 287 | 1 | 10 | 80 | 2 | 19 | 106 | 3 | 20 | 101 |
| 49 | 3 | 15 | 0 | 2 | 287 | 1 | 20 | 101 | 2 | 19 | 106 | 3 | 10 | 80 |
| 49 | 3 | 16 | 0 | 2 | 306 | 1 | 23 | 151 | 2 | 11 | 54 | 3 | 15 | 101 |
| 49 | 3 | 17 | 0 | 2 | 287 | 1 | 10 | 80 | 2 | 19 | 91 | 3 | 20 | 116 |
| 49 | 3 | 18 | 0 | 2 | 305 | 1 | 21 | 122 | 2 | 17 | 135 | 3 | 11 | 48 |
| 49 | 3 | 19 | 0 | 2 | 320 | 1 | 14 | 78 | 2 | 19 | 171 | 3 | 16 | 71 |
| 49 | 3 | 20 | 0 | 2 | 287 | 1 | 10 | 80 | 2 | 21 | 119 | 3 | 18 | 88 |

*Table 12.10*

```
BEST SOLUTION FOR 3 CLUSTERS AMONG 20 TRIALS

CLUSTER 1

  1   FTFTTFFFTFTTFFTFFFFTFFTTFFFTTTTFFTFFFTFTFFFFFFFFFFFFTFFFFT
  5   FTFTTTFFFFFTTFTTFTFTTFTTTTFTTTFFFTFFFTFTTTFFFFFFFFTFTTTFF
  7   FFTTFFFFTFFTFFTFFFTTFFFFFFFFTTFFFFFFFTFTFFFTFFFTTFFFFFFT
  9   FTFTFTFFTFFTFFTFFFTTFFFTFFFFTTFFFTFTFTTTFFFFFFFFFFFTTFFFT
 12   FTFTFFFFTFFTFFTFFFFTFFFTFFFFTTFTFTFFTTFFFFFFFFFFFFFFTFFTFF
 23   FTFTFFFTTFFTFFTFFTTTFFFFTFFFTTFTFTFFFTFFFTFFFFFFFFFFTFFFFT
 26   FTFFFFFFTFFTFFTFFFFFFFFFTFFFFTTFFFFFFFTFFTFFFFFFFFFFFFFFFF
 27   FTFTFFFFFFFTTFTFFFFTTFFTFFFFTTFTFTFFFTFFFFFFFFFFFFFFFTTFTFF
 35   TTFTFFFFTTFTFFTTFTTTFTFTTTFFTTFFTTFFFTFFFFFFFFFFFFFFFTFFFFF
 44   FTFTFTFFTFFTTFTFFFTTFFFTFFFFFFTTFFTFFFTFFFFFFFFFFFFFFTFFFFF

CLUSTER 2

  4   FFFFFTTTTFFTFFFFFFTTTFFFFFFFFFTFTFTFTFTTTFFFTTFFFFFFTTFFTF
  6   FFFFFFFFTFFTFFFFFFFFFFFFFFFFFTFFFFFTFTFTFFFFTFFFFFFFTFFFF
  8   TFTFFTFFTFFTFFFFFFTTFTFTFFFFFTFTFTFFFTFTFFFFFFFFFTFTTFFFF
 16   FFFFFTFFTFFTTFTFFFFFTFFFTFFFFFTFFFTFFFTFTFFFTFFFFFFFFTFFTF
 18   FFFFFTFFTFFTFFFFFFFTTFFFFFFFFFFTFTFTFTFTFTFFFFTFFFFFFFTFFTF
 19   FFFFFFFFTFFTFFFFTFTTFFFFFFFFFFTFFFTFFFTFTFFFFFFFFTFFTFFFF
 20   TFFFFFFFTFFTFFFFFFTTFTFFFFFFFFTFFFTFFFTFTFFFFFFFFFFFFFFFT
 21   TFFFFTFFTFFTFFFFTFTTFFFFFFFFFFTFFTTFFFTFTFFFTFFFFFFFTFFTF
 22   FFFFFTFFTFFTFFFFTFTTFFFTFFFFFTFTFTTTFTTTFTFFTFFFFFFFTFFTF
 24   FFTFFFFFTFFTFTFFFFFFTFFFTFFFFFFTFFFTFFFTFTFFFFTFFFFFFFTFFFF
 29   FFFFFTFFTFFTFFFFFFFTTFTFTFFTFFTFTTTTTFTFTFFFFFFFFFFFFFTFFFF
 30   FFFFFTFFTFFTFFFFFFFTTFFFFFFFFFFTFTFTFTFTFTFFFFFFFFFFFFFTFFTF
 32   FFFFFFFFFFFFTFTFFFFFFTFFFFFFFFFFTFTFTFTFTFTFFTFFTFTTFFTFFTF
 33   FFFFFTFFFFFFFFFFFFFFFFFFFFFFFFTFFFFFFFFFFFTFFFTTFFFFFFFTFFTF
 34   FFFFFFFFTFFTFFFFFFFFTFFFFFFFFFFTFFFTFTFTTTFFFFTFFFFFFFTFFFF
 36   FFFFFTFFTFFTFFFFTFTTTFFFFFFFFFTFTFTFFFTFTFFFFFTFFFTFFTFFTF
 37   FFFFFTFFFFFFTFFFFFFTTFFFFFFFFFFTTFFTFTFTFTFFFFFFFFFFFFFTFFFF
 41   FFFFFTFFTFFTFFFFFFFTFFFFFFFFFFFTFFFFFTFTTTFFFFTFFFTFFTFFFF
 42   FFFFFTFFTFFTFTTFFFFTTFFFTFFFFFFTFFFTFTFTFTFFFFTTFFFFFFTTFTF
 45   FFFFFTFFFFFFFFFFFFFFFFFFFFFFFFFFFFFFTFFFTFFFFTFFFFFFFTFFFF
 46   FFFFFFFTTFFTFFFFFFFTTFFFFFFFFFFTFFFTFTFTFTFFTFFFFFFTFFTFFFF
 48   FFFFFTFFTFFTFFFFFFFTTFFFFFFFFFFFTFFFTFTFTFTFFFFFFFFFFTFFTFFFF

CLUSTER 3

  2   FFFFFFFFFFFFFFFFFFFFFFFFFFFFFFFFFFFFFFFFFFFFFFFFFFFFTFFFFFFT
  3   FFFFFFFFFFFFFFFFFFFFFFFFFFFFFFFFFFFFFFFTFFFFFFFFFFFFFFTFFFFFFF
 10   FFFFFFFFTFFTFFTFFFFFFFFFFFFFFFFTFFFFFTFTFFFFTTFFFFFFFFFFFFT
 11   FFFFTFFFFFFTFFFFFFFTTFTFFFFTTTTFFTFFFTFFFFFFFFFFFFFFFFFFFF
 13   FTFTFFFFFFFTTFTFFFFFFFFTFFFFTTFFFFFTFTFFTFFFFFFTFFFFFFFFF
 14   FFFFFFFFFFFTFFTFFFFFFFFFFFFFTTFFFFFTFTFFFFFTFFFFFFFFFFFT
 15   FFFFFFFFTFFTFFFFFFFFFFFFFFFFFTFFFFFTFTFTFFFFFFFFFFFFFFFFF
 17   FFFTFFFFFFFTFFFFFFFFFFFFFFFFTTFFFFFTFTFFFFFFFFFFFFFFFFFFT
 25   TFFFFFFFTFFFFFFFFFFFFFTFFFTFFFFFFFFFFFFFFFFFFFFFFFFFFFFFFT
 28   FTFTFFFFFFFFFFFFFFFFFFFFFFFFTTFFFFFFFTFFFFFTFFFFFFFFFFFFF
 31   FFFFFFFFTFFTFFFFFFFFFFFFFFFFFTFFFFFFFTFTFFFFFFTFFFFFFFFF
 38   FFFFFFFFTFFTFFFFFFFFFFFFFFFFFTFFFFFTFTFFTFFTFFFFFFFFFFFT
 39   FFFFFFFFFFFFFFFFFFFFFFFFFFFFFTFFFFFTFTFFFFFTFFFFFFFFFFFF
 40   FFFFFFFFFFFTFFTFFFFFFFFFFFFFTTFFFFFTFTFFFFFFFFFFFFFFFFFFT
 43   FFFTFFFFFFFTFFTFFFFFFFFFFFFFTTFFFFFTFTFFFFFFFFFFFFFFFFFFFT
 47   FFFFFFFFTFFTFFFFFFFFFFFFFFFFFTFFFFFFFTFFFFFFFFFFFFFFFFFFF
 49   FFFTFFFFFFFTFFTFFFFFFFFFFFFFTTFFFFFTFTFFFFFTFFFFFFFFFFFT
```

*Table 12.11*

```
49 6  1 0 4 235    1  9   29 2 12   61 3  6   25 4  7   35 5  6   16 6  9   69
49 6  2 0 2 249    1 15   69 2  7   26 3  3   25 4  9   60 5  6   22 6  9   47
49 6  3 0 2 252    1  8   64 2 11   40 3  9   44 4  9   41 5  5   26 6  7   37
49 6  4 0 4 239    1  9   45 2  3   13 3 11   48 4  9   69 5  9   29 6  8   35
49 6  5 0 3 243    1  8   23 2 14   76 3  7   44 4 10   44 5  3   23 6  7   33
49 6  6 0 3 234    1  8   38 2  7   24 3 11   53 4  9   29 5  9   69 6  5   21
49 6  7 0 3 240    1  8   36 2  8   64 3  9   29 4  7   29 5  7   37 6 10   45
49 6  8 0 4 236    1  3   10 2  9   69 3  8   25 4  9   31 5  9   44 6 11   57
49 6  9 0 2 238    1  7   29 2 10   34 3  8   35 4  7   31 5  9   69 6  8   40
49 6 10 0 3 241    1  9   69 2  8   37 3  7   29 4  7   43 5  9   34 6  9   29
49 6 11 0 3 239    1  9   69 2  8   25 3  3   13 4 12   56 5 10   47 6  7   29
49 6 12 0 3 247    1  6   22 2  9   69 3  9   46 4 13   56 5  5   28 6  7   26
49 6 13 0 2 256    1 11   37 2 16   84 3  9   74 4  5   26 5  5   21 6  3   14
49 6 14 0 2 242    1  9   69 2  3    6 3  7   24 4  7   36 5 16   88 6  7   19
49 6 15 0 2 243    1 11   48 2  9   69 3  7   16 4  7   37 5  6   27 6  9   46
49 6 16 0 3 246    1  8   64 2  5   26 3  7   25 4 10   49 5  7   31 6 12   51
49 6 17 0 3 238    1 11   57 2  8   64 3  7   15 4  5   18 5 11   51 6  7   33
49 6 18 0 3 245    1  5   24 2  9   69 3  6   34 4  8   25 5 11   46 6 10   47
49 6 19 0 3 241    1 11   46 2 10   54 3  9   69 4  5   17 5  8   25 6  6   30
49 6 20 0 2 247    1  8   36 2  9   69 3 13   56 4  6   30 5  7   35 6  6   21
```

*Table 12.12*

```
BEST SOLUTION FOR 6 CLUSTERS AMONG 20 TRIALS

CLUSTER 1

  4     FFFFFTTTTFFTFFFFFFFTTTFFFFFFFFTFTFTFTFTTTFFFFTTFFFFFFTTFFTF
 18     FFFFFTFFTFFTFFFFFFFTTFFFFFFFFFTFTFTFTFTFTFFFFTFFFFFFFFTFFTF
 22     FFFFFTFFTFFTFFFFTFTTFFFTFFFFFFTFTFTTTFTTTFTFFTFFFFFFFFTFFTF
 29     FFFFFTFFTFFTFFFFFFFTTFTFTFFTFFTFTTTTTFTFTFFFFFFFFFFFFFTFFFF
 30     FFFFFTFFTFFTFFFFFFFTTFFFFFFFFFTFTFTFTFTFTFFFFFFFFFFFFFTFFTF
 32     FFFFFFFFFFFTFTFFFFFFTFFFFFFFFFTFTFTFTFTFFTFFTFTTFFTFFTF
 36     FFFFFTFFTFFTFFFFTFTTTFFFFFFFFFTFTFTFFFTFTFFFFTFFFFTFFTFFTF
 42     FFFFFTFFTFFTFTTFFFFTTFFFTFFFFFFTFFFTFTFTFTFFFFFTTFFFFFFTTFTF

CLUSTER 2

  6     FFFFFFFFTFFTFFFFFFFFFFFFFFFFFFTFFFFFFTFTFTFFFFTFFFFFFFTFFFF
 11     FFFFTFFFFFFTFFFFFFFFTTFTFFFFTTTFFFTFFFFTFFFFFFFFFFFFFFFFFFF
 15     FFFFFFFFTFFTFFFFFFFFFFFFFFFFFFTFFFFFFTFTFTFFFFFFFFFFFFFFFFF
 31     FFFFFFFFTFFTFFFFFFFFFFFFFFFFFFTFFFFFFFTFTFFFFFFTFFFFFFFFFFF
 38     FFFFFFFFTFFTFFFFFFFFFFFFFFFFFFTFFFFFFTFTFFTFFTFFFFFFFFFFFFT
 39     FFFFFFFFFFFFFFFFFFFFFFFFFFFFFFTFFFFFFTFTFFFFFFTFFFFFFFFFFFF
 47     FFFFFFFFTFFTFFFFFFFFFFFFFFFFFFTFFFFFFFFTFFFFFFFFFFFFFFFFFFF

CLUSTER 3

  8     TFTFFTFFTFFTFFFFFFFTTFTFTFFFFFTFTFTFFFTFTFFFFFFFFFFTFTTFFFF
 16     FFFFFTFFTFFTTFTFFFFTFFFTFFFFFFTFFFTFFFTFTFFFTFFFFFFFFFTFFTF
 19     FFFFFFFFTFFTFFFFTFTTFFFFFFFFFFTFFFTFFFTFTFFFFFFFFFFTFFTFFFF
 20     TFFFFFFFTFFTFFFFFFFTTFTFFFFFFFTFFFTFFFTFTFFFFFFFFFFFFFFFFFT
 21     TFFFFTFFTFFTFFFFTFTTFFFFFFFFFFTFFTTFFFTFTFFFTFFFFFFFFFTFFTF
 24     FFTFFFFFTFFTFTFFFFFTFFFTFFFFFFTFFFTFFFTFTFFFFTFFFFFFFFTFFFF
 34     FFFFFFFFTFFTFFFFFFFTFFFFFFFFFFTFFFTFTFTTTFFFFTFFFFFFFFTFFFF
 37     FFFFFTFFFFFTFFFFFFFTTFFFFFFFFFTTFFTFTFTFTFFFFFFFFFFFFFTFFFF
 41     FFFFFTFFTFFTFFFFFFFTFFFFFFFFFFTFFFFFTFTTTFFFFTFFFFFTFFTFFFF
 46     FFFFFFFTTFFTFFFFFFFTTFFFFFFFFFTFFFTFTFTFTFFTFFFFFFFTFFTFFFF
 48     FFFFFTFFTFFTFFFFFFFTTFFFFFFFFFTFFFTFTFTFTFFFFFFFFFFTFFTFFFF
```

*Table 12.12 – continued*

```
CLUSTER 4

  7   FFTTFFFFTFFTFFTFFFTTFFFFFFFFTTFFFFFFFTFTFFFTFFFTTFFFFFFT
 10   FFFFFFFFTFFTFFTFFFFFFFFFFFFFFTFFFFFTFTFFFFTTFFFFFFFFFFFT
 13   FTFTFFFFFFFTTFTFFFFFFFFTFFFFTTFFFFFTFTFFTFFFFFTFFFFFFFFT
 14   FFFFFFFFFFFTFFTFFFFFFFFFFFFFTTFFFFFTFTFFFFFTFFFFFFFFFFFT
 17   FFFTFFFFFFFTFFFFFFFFFFFFFFFFTTFFFFFTFTFFFFFFFFFFFFFFFFFT
 28   FTFTFFFFFFFFFFFFFFFFFFFFFFFFTTFFFFFFFTFFFFFTFFFFFFFFFFFF
 40   FFFFFFFFFFFTFFTFFFFFFFFFFFFFTTFFFFFTFTFFFFFFFFFFFFFFFFFT
 43   FFFTFFFFFFFTFFTFFFFFFFFFFFFFTTFFFFFTFTFFFFFFFFFFFFFFFFFT
 49   FFFTFFFFFFFTFFTFFFFFFFFFFFFFTTFFFFFTFTFFFFFTFFFFFFFFFFFT

CLUSTER 5

  1   FTFTTFFFTFTTFFTFFFFTFFTTFFFTTTTFFTFFFTFTFFFFFFFFFFTFFFFT
  5   FTFTTTFFFFFTTFTTFTFTTFTTTTFTTTFFFTFFFTFTTTFFFFFFFTFTTTFF
  9   FTFTFTFFTFFTFFTFFFTTFFFTFFFFTTFFFTFTFTTTFFFFFFFFFFTTFFFT
 12   FTFTFFFFTFFTFFTFFFFTFFFTFFFFTTFTFTFFTTFFFFFFFFFFFFTFFTFF
 23   FTFTFFFTTFFTFFTFFTTTFFFFTFFFTTFTFTFFFTFFFTFFFFFFFFTFFFFT
 26   FTFFFFFFTFFTFFTFFFFFFFFTFFFFTTFFFFFFFTFFTFFFFFFFFFFFFFFF
 27   FTFTFFFFFFFTTFTFFFFTTFFTFFFFTTFTFTFFFTFFFFFFFFFFFFTTFTFF
 35   TTFTFFFFTTFTFFTTFTTTFTFTTTFFTTFFTTFFFTFFFFFFFFFFFFTFFFFF
 44   FTFTFTFFTFFTTFTFFFTTFFFTFFFFFTTFFTFFFTFFFFFFFFFFFFTFFFFF

CLUSTER 6

  2   FFFFFFFFFFFFFFFFFFFFFFFFFFFFFFFFFFFFFFFFFFFFFFFFTFFFFFFT
  3   FFFFFFFFFFFFFFFFFFFFFFFFFFFFFFFFFFFTFFFFFFFFFFFFTFFFFFFF
 25   TFFFFFFFTFFFFFFFFFFFFTFFFTFFFFFFFFFFFFFFFFFFFFFFFFFFFFFT
 33   FFFFFTFFFFFFFFFFFFFFFFFFFFFFFTFFFFFFFFFTFFFTTFFFFFFTFFTF
 45   FFFFFTFFFFFFFFFFFFFFFFFFFFFFFFFFFFFTFFFTFFFFTFFFFFFTFFFF
```

## 12.3 Sample Main Program and Two Test Examples for Three Centre-Free Criteria

For the test of TIHEXM we provide two examples. One is based on a distance matrix for m = 11, the other on one for m = 42. Owing to space limitations, we cannot deal with bigger numbers of objects here. However, despite a storage requirement of $m(m-1)/2 + 3n + m$ for TIHEXM, there are successful applications, for example with $m \approx 1000$ on an IBM 370/158 [B8].

The example with $m = 11$ objects deals with 11 makes of car. Associated with each possible pair (i, h) ($i = 2, \ldots, m$, $h = 1 \ldots, i-1$) is one of the numbers $1, \ldots, m(m-1)/2 = 55$. In other words, a ranking sequence of the pairs is formed [A20]. Using the main program in Table 12.13, the corresponding matrix $(t_{ih})$ is printed out in Table 12.14. The main program expects to read the data as integers, as a continuous stream of rows, using the FORMAT (16I5). These are then transferred to the real array T. DO-loop number 2 shows how (10.10) may be used to print the linearly stored data in a suitable way. Apart from this, the main program is structured like the others.

*Table 12.13*

```
C   SAMPLE MAIN PROGRAM FOR TIHEXM
C
C
C   THREE METHODS
C
C   M=200 (MAXIMUM)
C   M12 = (M*(M-1))/2 ( = 19900)
C   N=9 (MAXIMUM)
C
      INTEGER  KT(19900),Z(200),P(200),MJ(9),H
      DIMENSION T(19900),E(9),C(9)
      KI=5
      KO=6
C
      M=11
      N1=2
      N2=3
      NRMAX=10
C
      BIG=1.E50
      IY=37519281
      M12=(M*(M-1))/2
      READ(KI,28) (KT(K),K=1,M12)
      DO 1 K=1,M12
          T(K)=KT(K)
    1 CONTINUE
      WRITE(KO,20)
      WRITE(KO,21) (I,I=1,M)
      K=1
      L=1
      DO 2 I=2,M
          WRITE(KO,22) I,(KT(H),H=K,L)
          K=K+I-1
          L=L+I
    2 CONTINUE
      DO 6 N=N1,N2
          WRITE(KO,24)
          DO 5 METHOD=1,3
          WRITE(KO,23) METHOD
          WRITE(KO,21)
          IY=37519281
          DD=BIG
          DO 4 NR=1,NRMAX
              CALL RANDP (M,N,Z,IY)
              CALL TIHEXM (M,M12,T,N,MJ,METHOD,
     *                         Z,E,D,IT,IFLAG,C)
              WRITE(KO,25) M,N,NR,IFLAG,IT,D,
     *                         (J,MJ(J),E(J),J=1,N)
              IF(IFLAG.NE.0) GOTO 4
              IF(D.GE.DD) GOTO 4
              DD=D
              DO 3 I=1,M
                  P(I)=Z(I)
    3         CONTINUE
    4     CONTINUE
          WRITE(KO,26) (P(I),I=1,M)
          WRITE(KO,21)
    5     CONTINUE
    6 CONTINUE
   20 FORMAT('1',4X,'GIVEN MATRIX T(I,H)')
   21 FORMAT('0',4X,42I3/(5X,42I3))
   22 FORMAT(' ',I3,1X,42I3/(5X,42I3))
   23 FORMAT(' ','METHOD = ',I1)
   24 FORMAT('1')
```

*Table 12.13 – continued*

```
      25 FORMAT(' ',I2,1X,I1,1X,I2,I2,I2,F8.1,3X,9(I2,I3,F7.1))
      26 FORMAT('0','BEST SOLUTION: ',50I1/(16X,50I1))
      27 FORMAT('1')
      28 FORMAT(16I5)
         STOP
         END
```

*Table 12.14*

GIVEN MATRIX T(I,H)

| | 1 | 2 | 3 | 4 | 5 | 6 | 7 | 8 | 9 | 10 | 11 |
|---|---|---|---|---|---|---|---|---|---|---|---|
| 2 | 8 | | | | | | | | | | |
| 3 | 50 | 38 | | | | | | | | | |
| 4 | 31 | 9 | 11 | | | | | | | | |
| 5 | 12 | 33 | 55 | 44 | | | | | | | |
| 6 | 48 | 37 | 1 | 13 | 54 | | | | | | |
| 7 | 36 | 22 | 23 | 16 | 53 | 26 | | | | | |
| 8 | 2 | 6 | 46 | 19 | 30 | 47 | 29 | | | | |
| 9 | 5 | 4 | 41 | 25 | 28 | 40 | 35 | 3 | | | |
| 10 | 39 | 14 | 17 | 18 | 45 | 24 | 34 | 27 | 20 | | |
| 11 | 10 | 32 | 52 | 42 | 7 | 51 | 49 | 15 | 21 | 43 | |

Tables 12.15 and 12.16 show the results of applying TIHEXM for N = 2 and N = 3, with NRMAX = 10 random starting partitions and using the three objective functions determined by the parameter METHOD. For NR = 1, NRMAX, the rows of the tables show the values of M, N, NR, IFLAG, IT, D, (J, MJ(J), E(J), J=1, N). The variables have their usual meaning. Finally, the tables show the vector of assignments (Z(I), I=1, M) for the (first) best solution. For N = 2, the best solutions found are the same for all three methods. However, with METHOD = 1, the best solution occurs 9 times, with METHOD = 2 even 10 times, but with METHOD = 3 only 3 times. For N = 3, the best solutions differ. With METHOD = 3, the case IFLAG = 2 occurs once, and the other nine times the clusters are formed in such a way that only one cluster has more than two elements. We have previously indicated this possibility. The computation time on the TR 440 amounted to three seconds.

In the second example, with M = 42 objects, we have generated the distance matrix using a modified version of the main program. The change, shown in Table 12.17, involves only the beginning of the program. The first DO-loop generates the distance matrix which is printed in Table 12.18. Because of the properties of URAND, this may be different on different computers. Tables 12.19, 12.20 and 12.21 show the results from this example for N = 2, 3, 4 and all three methods. The format is the same as that for the first example.

It is noticeable that, with METHOD = 1 and METHOD = 2, many different values of the objective function occur, whatever the number of classes. The explanation is, of course, that we cannot expect a structure in the set of objects, given that randomly generated distances were used. With METHOD = 3 this expresses itself in the fact that the best solutions found contain only one cluster with more than two elements. The computation time for this second example was 64 seconds on the TR 440.

Whereas METHOD = 1 and METHOD = 2 may both be considered as useful alternatives, caution is necessary when using METHOD = 3.

*Table 12.15*

```
METHOD = 1

11 2  1 0 3    399.0    1  6   216.0 2  5   183.0
11 2  2 0 3    399.0    1  5   183.0 2  6   216.0
11 2  3 0 1    399.0    1  6   216.0 2  5   183.0
11 2  4 0 2    399.0    1  6   216.0 2  5   183.0
11 2  5 0 1    614.0    1  5   330.0 2  6   284.0
11 2  6 0 3    399.0    1  5   183.0 2  6   216.0
11 2  7 0 3    399.0    1  5   183.0 2  6   216.0
11 2  8 0 2    399.0    1  6   216.0 2  5   183.0
11 2  9 0 1    399.0    1  5   183.0 2  6   216.0
11 2 10 0 2    399.0    1  5   183.0 2  6   216.0

BEST SOLUTION: 11221221121

METHOD = 2

11 2  1 0 2     72.6    1  6    36.0 2  5    36.6
11 2  2 0 2     72.6    1  5    36.6 2  6    36.0
11 2  3 0 1     72.6    1  6    36.0 2  5    36.6
11 2  4 0 2     72.6    1  5    36.6 2  6    36.0
11 2  5 0 2     72.6    1  5    36.6 2  6    36.0
11 2  6 0 2     72.6    1  5    36.6 2  6    36.0
11 2  7 0 2     72.6    1  5    36.6 2  6    36.0
11 2  8 0 2     72.6    1  6    36.0 2  5    36.6
11 2  9 0 2     72.6    1  5    36.6 2  6    36.0
11 2 10 0 2     72.6    1  5    36.6 2  6    36.0

BEST SOLUTION: 11221221121

METHOD = 3

11 2  1 0 2     16.4    1  6     7.2 2  5     9.2
11 2  2 0 1     21.9    1  2     7.0 2  9    14.9
11 2  3 0 1     16.4    1  6     7.2 2  5     9.2
11 2  4 0 2     16.7    1  9    15.2 2  2     1.5
11 2  5 0 2     16.3    1  5     9.2 2  6     7.2
11 2  6 0 2     16.7    1  9    15.2 2  2     1.5
11 2  7 0 1     16.3    1  6     7.2 2  5     9.2
11 2  8 0 2     16.4    1  6     7.2 2  5     9.2
11 2  9 0 2     16.3    1  5     9.2 2  6     7.2
11 2 10 0 2     16.7    1  9    15.2 2  2     1.5

BEST SOLUTION: 22112112212
```

*Table 12.16*

```
METHOD = 1


11 3  1 0 1   222.0   1  3   42.0 2  5  133.0 3  3   47.0
11 3  2 0 3   193.0   1  4   74.0 2  4   90.0 3  3   29.0
11 3  3 0 2   212.0   1  3   29.0 2  4   84.0 3  4   99.0
11 3  4 0 5   193.0   1  4   74.0 2  3   29.0 3  4   90.0
11 3  5 0 3   213.0   1  4   90.0 2  3   95.0 3  4   28.0
11 3  6 0 2   213.0   1  3   95.0 2  4   28.0 3  4   90.0
11 3  7 0 2   213.0   1  3   95.0 2  4   28.0 3  4   90.0
11 3  8 0 1   222.0   1  5  133.0 2  3   42.0 3  3   47.0
11 3  9 0 2   213.0   1  4   90.0 2  4   28.0 3  3   95.0
11 3 10 0 2   228.0   1  3   25.0 2  3   70.0 3  5  133.0

BEST SOLUTION: 31223221113

METHOD = 2


11 3  1 0 2    47.1   1  4    7.0 2  2    3.5 3  5   36.6
11 3  2 0 2    47.1   1  4    7.0 2  5   36.6 3  2    3.5
11 3  3 0 2    47.1   1  2    3.5 2  4    7.0 3  5   36.6
11 3  4 0 2    57.0   1  1    0.0 2  6   36.0 3  4   21.0
11 3  5 0 4    47.1   1  5   36.6 2  4    7.0 3  2    3.5
11 3  6 0 2    55.4   1  2    0.5 2  5   26.6 3  4   28.3
11 3  7 0 1    47.1   1  2    3.5 2  4    7.0 3  5   36.6
11 3  8 0 2    47.1   1  2    3.5 2  5   36.6 3  4    7.0
11 3  9 0 3    47.1   1  5   36.6 2  4    7.0 3  2    3.5
11 3 10 0 1    47.1   1  2    3.5 2  5   36.6 3  4    7.0

BEST SOLUTION: 22331332231

METHOD = 3


11 3  1 0 2    13.6   1  7    9.6 2  2    3.5 3  2    0.5
11 3  2 0 1    19.1   1  2    2.0 2  7   16.1 3  2    1.0
11 3  3 0 2    22.1   1  2    3.5 2  7   14.1 3  2    4.5
11 3  4 0 3    15.0   1  5    9.2 2  4    2.3 3  2    3.5
11 3  5 0 1    21.6   1  7   16.1 2  2    4.0 3  2    1.5
11 3  6 2 0     0.0   1  6    0.0 2  1    0.0 3  4    0.0
11 3  7 0 3    15.0   1  5    9.2 2  4    2.3 3  2    3.5
11 3  8 0 1    15.0   1  2    3.5 2  5    9.2 3  4    2.3
11 3  9 0 2    13.6   1  2    0.5 2  7    9.6 3  2    3.5
11 3 10 0 2    15.0   1  2    3.5 2  5    9.2 3  4    2.3

BEST SOLUTION: 11312311112
```

*Table 12.17*

```
C   SAMPLE MAIN PROGRAM FOR TIHEXM
C
C
C   THREE METHODS
C
C   M=200 (MAXIMUM)
C   M12 = (M*(M-1))/2 ( = 19900)
C   N=9 (MAXIMUM)
C
      INTEGER  KT(19900),Z(200),P(200),MJ(9),H
      DIMENSION T(19900),E(9),C(9)
      KO=6
C
      M=42
      N1=2
      N2=4
      NRMAX=10
C
      BIG=1.E50
      IY=37519281
      M12=(M*(M-1))/2
      DO 1 K=1,M12
            F=URAND(IY)
            IF(F.LE..3) G=F*15.
            IF(F.GT..3) G=F*99.9
            L=IFIX(G)
            KT(K)=L
            T(K)=L
    1 CONTINUE
      .
      .
      .
      STOP
      END
```

GIVEN MATRIX T(I,H)

| | 1 | 2 | 3 | 4 | 5 | 6 | 7 | 8 | 9 | 10 | 11 | 12 | 13 | 14 | 15 | 16 | 17 | 18 | 19 | 20 | 21 | 22 | 23 | 24 | 25 | 26 | 27 | 28 | 29 | 30 | 31 | 32 | 33 | 34 | 35 | 36 | 37 | 38 | 39 | 40 | 41 | 42 |
|---|---|---|---|---|---|---|---|---|---|---|---|---|---|---|---|---|---|---|---|---|---|---|---|---|---|---|---|---|---|---|---|---|---|---|---|---|---|---|---|---|---|---|
| 2 | 40 | | | | | | | | | | | | | | | | | | | | | | | | | | | | | | | | | | | | | | | | |
| 3 | 33 | 31 | | | | | | | | | | | | | | | | | | | | | | | | | | | | | | | | | | | | | | | |
| 4 | 85 | 60 | 68 | | | | | | | | | | | | | | | | | | | | | | | | | | | | | | | | | | | | | | |
| 5 | 4 | 50 | 4 | 75 | | | | | | | | | | | | | | | | | | | | | | | | | | | | | | | | | | | | | |
| 6 | 43 | 2 | 3 | 86 | 0 | | | | | | | | | | | | | | | | | | | | | | | | | | | | | | | | | | | | |
| 7 | 70 | 56 | 64 | 93 | 0 | 3 | | | | | | | | | | | | | | | | | | | | | | | | | | | | | | | | | | | |
| 8 | 62 | 71 | 3 | 61 | 64 | 93 | 80 | | | | | | | | | | | | | | | | | | | | | | | | | | | | | | | | | | |
| 9 | 35 | 99 | 36 | 45 | 31 | 1 | 50 | 0 | | | | | | | | | | | | | | | | | | | | | | | | | | | | | | | | | |
| 10 | 34 | 68 | 3 | 0 | 0 | 34 | 73 | 3 | 89 | | | | | | | | | | | | | | | | | | | | | | | | | | | | | | | | |
| 11 | 38 | 0 | 2 | 54 | 1 | 90 | 93 | 87 | 58 | 2 | | | | | | | | | | | | | | | | | | | | | | | | | | | | | | | |
| 12 | 1 | 2 | 2 | 1 | 98 | 96 | 80 | 86 | 64 | 0 | 0 | | | | | | | | | | | | | | | | | | | | | | | | | | | | | | |
| 13 | 2 | 1 | 0 | 65 | 82 | 70 | 0 | 73 | 51 | 76 | 3 | 30 | | | | | | | | | | | | | | | | | | | | | | | | | | | | | |
| 14 | 36 | 52 | 47 | 4 | 39 | 71 | 99 | 39 | 54 | 0 | 89 | 3 | 85 | | | | | | | | | | | | | | | | | | | | | | | | | | | | |
| 15 | 2 | 0 | 1 | 1 | 49 | 82 | 54 | 98 | 0 | 1 | 57 | 37 | 89 | 34 | | | | | | | | | | | | | | | | | | | | | | | | | | | |
| 16 | 46 | 62 | 70 | 31 | 2 | 48 | 0 | 43 | 1 | 66 | 1 | 60 | 98 | 53 | 0 | | | | | | | | | | | | | | | | | | | | | | | | | | |
| 17 | 61 | 2 | 3 | 85 | 0 | 0 | 0 | 59 | 69 | 94 | 87 | 65 | 50 | 91 | 60 | 60 | | | | | | | | | | | | | | | | | | | | | | | | | |
| 18 | 79 | 55 | 98 | 0 | 60 | 0 | 3 | 3 | 2 | 50 | 2 | 59 | 73 | 75 | 2 | 0 | 0 | | | | | | | | | | | | | | | | | | | | | | | | |
| 19 | 40 | 45 | 93 | 90 | 64 | 0 | 86 | 95 | 96 | 80 | 91 | 3 | 67 | 43 | 90 | 94 | 0 | 31 | | | | | | | | | | | | | | | | | | | | | | | |
| 20 | 81 | 2 | 65 | 43 | 3 | 98 | 2 | 0 | 50 | 90 | 63 | 3 | 92 | 50 | 45 | 54 | 40 | 89 | 30 | | | | | | | | | | | | | | | | | | | | | | |
| 21 | 1 | 3 | 74 | 68 | 83 | 57 | 3 | 98 | 2 | 69 | 99 | 73 | 2 | 45 | 3 | 0 | 98 | 2 | 1 | 77 | | | | | | | | | | | | | | | | | | | | | |
| 22 | 66 | 2 | 73 | 65 | 1 | 35 | 81 | 99 | 3 | 54 | 58 | 61 | 61 | 96 | 90 | 93 | 84 | 2 | 39 | 3 | 4 | | | | | | | | | | | | | | | | | | | | |
| 23 | 87 | 70 | 52 | 33 | 1 | 0 | 57 | 49 | 45 | 60 | 4 | 4 | 38 | 86 | 2 | 87 | 71 | 76 | 52 | 34 | 2 | 57 | | | | | | | | | | | | | | | | | | | |
| 24 | 0 | 0 | 40 | 3 | 61 | 1 | 59 | 64 | 90 | 1 | 97 | 32 | 65 | 63 | 2 | 53 | 2 | 51 | 73 | 56 | 76 | 3 | 85 | | | | | | | | | | | | | | | | | | |
| 25 | 1 | 60 | 64 | 1 | 77 | 2 | 48 | 99 | 79 | 70 | 73 | 1 | 44 | 79 | 86 | 85 | 70 | 48 | 54 | 2 | 43 | 51 | 0 | 32 | | | | | | | | | | | | | | | | | |
| 26 | 3 | 1 | 65 | 2 | 0 | 2 | 93 | 0 | 71 | 57 | 2 | 81 | 59 | 69 | 2 | 32 | 66 | 3 | 2 | 1 | 32 | 58 | 54 | 78 | 67 | | | | | | | | | | | | | | | | |
| 27 | 60 | 2 | 76 | 86 | 63 | 1 | 0 | 93 | 3 | 86 | 3 | 40 | 82 | 3 | 1 | 2 | 83 | 67 | 3 | 87 | 2 | 0 | 70 | 88 | 76 | 74 | | | | | | | | | | | | | | | |
| 28 | 37 | 65 | 85 | 0 | 97 | 99 | 2 | 2 | 0 | 51 | 1 | 92 | 1 | 2 | 2 | 78 | 90 | 91 | 88 | 83 | 81 | 69 | 58 | 35 | 31 | 75 | 43 | | | | | | | | | | | | | | |
| 29 | 87 | 95 | 0 | 57 | 95 | 67 | 42 | 65 | 81 | 2 | 70 | 4 | 37 | 53 | 2 | 94 | 32 | 59 | 51 | 48 | 3 | 4 | 3 | 32 | 60 | 0 | 74 | 36 | | | | | | | | | | | | | |
| 30 | 2 | 1 | 45 | 4 | 70 | 3 | 99 | 33 | 86 | 77 | 60 | 54 | 62 | 30 | 84 | 66 | 80 | 85 | 46 | 41 | 31 | 42 | 3 | 81 | 88 | 43 | 0 | 95 | 56 | | | | | | | | | | | | |
| 31 | 95 | 95 | 49 | 45 | 0 | 85 | 70 | 2 | 31 | 74 | 0 | 42 | 34 | 43 | 67 | 50 | 34 | 1 | 69 | 71 | 72 | 43 | 85 | 2 | 0 | 47 | 35 | 83 | 0 | 4 | | | | | | | | | | | |
| 32 | 1 | 89 | 80 | 82 | 1 | 1 | 72 | 0 | 49 | 39 | 1 | 40 | 0 | 92 | 93 | 1 | 60 | 4 | 66 | 79 | 78 | 37 | 56 | 55 | 39 | 55 | 97 | 92 | 1 | 54 | 1 | | | | | | | | | | |
| 33 | 98 | 2 | 74 | 50 | 49 | 68 | 38 | 64 | 76 | 3 | 36 | 1 | 59 | 3 | 72 | 69 | 95 | 0 | 1 | 2 | 89 | 60 | 37 | 54 | 38 | 69 | 35 | 3 | 70 | 4 | 69 | 3 | | | | | | | | | |
| 34 | 36 | 93 | 91 | 80 | 70 | 1 | 3 | 81 | 1 | 30 | 36 | 41 | 76 | 2 | 89 | 3 | 32 | 3 | 87 | 1 | 49 | 56 | 57 | 90 | 37 | 0 | 53 | 87 | 61 | 74 | 4 | 56 | 57 | | | | | | | | |
| 35 | 59 | 35 | 2 | 0 | 3 | 33 | 61 | 90 | 69 | 92 | 61 | 1 | 4 | 1 | 1 | 52 | 4 | 3 | 37 | 37 | 0 | 64 | 85 | 58 | 86 | 56 | 84 | 3 | 43 | 64 | 57 | 87 | 82 | 58 | | | | | | | |
| 36 | 3 | 49 | 4 | 70 | 0 | 2 | 52 | 2 | 85 | 39 | 1 | 56 | 63 | 67 | 3 | 89 | 89 | 83 | 2 | 4 | 78 | 42 | 3 | 2 | 1 | 85 | 99 | 87 | 76 | 4 | 36 | 97 | 3 | 0 | 41 | | | | | | |
| 37 | 90 | 43 | 62 | 3 | 39 | 3 | 3 | 66 | 0 | 2 | 77 | 79 | 43 | 2 | 80 | 78 | 94 | 95 | 39 | 92 | 3 | 31 | 87 | 31 | 1 | 72 | 70 | 56 | 78 | 48 | 53 | 0 | 90 | 65 | 3 | 1 | | | | | |
| 38 | 2 | 3 | 55 | 73 | 2 | 76 | 38 | 2 | 1 | 88 | 1 | 66 | 3 | 57 | 72 | 89 | 66 | 88 | 4 | 1 | 83 | 30 | 66 | 71 | 2 | 40 | 3 | 57 | 45 | 59 | 49 | 65 | 71 | 44 | 2 | 33 | 40 | | | | |
| 39 | 3 | 70 | 47 | 45 | 94 | 0 | 52 | 80 | 37 | 1 | 60 | 65 | 72 | 60 | 64 | 87 | 68 | 4 | 1 | 1 | 36 | 77 | 96 | 2 | 2 | 98 | 89 | 46 | 34 | 72 | 0 | 87 | 72 | 3 | 76 | 96 | 40 | 2 | | | |
| 40 | 1 | 3 | 47 | 37 | 0 | 3 | 50 | 68 | 43 | 67 | 32 | 95 | 84 | 83 | 79 | 1 | 45 | 4 | 70 | 69 | 51 | 37 | 4 | 96 | 68 | 47 | 69 | 87 | 39 | 3 | 83 | 4 | 40 | 85 | 78 | 1 | 83 | 67 | 98 | | |
| 41 | 3 | 0 | 30 | 90 | 44 | 81 | 68 | 64 | 2 | 2 | 61 | 80 | 45 | 43 | 3 | 69 | 86 | 61 | 67 | 1 | 39 | 64 | 36 | 54 | 62 | 0 | 93 | 79 | 2 | 84 | 30 | 51 | 98 | 72 | 32 | 2 | 41 | 86 | 31 | 69 | |
| 42 | 3 | 94 | 83 | 54 | 31 | 72 | 59 | 45 | 53 | 52 | 2 | 56 | 61 | 45 | 94 | 40 | 50 | 4 | 85 | 74 | 69 | 82 | 84 | 73 | 80 | 62 | 82 | 50 | 49 | 3 | 0 | 39 | 1 | 2 | 67 | 2 | 89 | 75 | 75 | 31 | 74 |

*Table 12.18*

*Table 12.19*

```
METHOD = 1


42 2  1 0 5 16254.0     1 21 8486.0 2 21 7768.0
42 2  2 0 3 16306.0     1 21 7764.0 2 21 8542.0
42 2  3 0 5 16741.0     1 21 8229.0 2 21 8512.0
42 2  4 0 5 16145.0     1 21 8273.0 2 21 7872.0
42 2  5 0 4 16198.0     1 22 8691.0 2 20 7507.0
42 2  6 0 3 16252.0     1 21 7714.0 2 21 8538.0
42 2  7 0 4 16091.0     1 20 7511.0 2 22 8580.0
42 2  8 0 2 16256.0     1 21 7806.0 2 21 8450.0
42 2  9 0 4 16126.0     1 21 8318.0 2 21 7808.0
42 2 10 0 5 16512.0     1 21 7922.0 2 21 8590.0

BEST SOLUTION: 221122212111112222222211122111111221122221

METHOD = 2


42 2  1 0 4    763.8     1 24  474.9 2 18  288.9
42 2  2 0 4    749.6     1 18  294.7 2 24  454.9
42 2  3 0 3    775.5     1 23  429.7 2 19  345.7
42 2  4 0 2    777.2     1 24  472.4 2 18  304.8
42 2  5 0 5    752.0     1 27  536.3 2 15  215.7
42 2  6 0 5    760.5     1 21  383.0 2 21  377.6
42 2  7 0 4    749.6     1 24  454.9 2 18  294.7
42 2  8 0 4    760.9     1 22  399.3 2 20  361.6
42 2  9 0 3    768.3     1 23  412.2 2 19  356.1
42 2 10 0 2    769.7     1 24  461.5 2 18  308.2

BEST SOLUTION: 222211111212222111112112211221111121212121

METHOD = 3


42 2  1 0 5      22.3     1 40    22.3 2  2      0.0
42 2  2 0 4      22.9     1 40    22.9 2  2      0.0
42 2  3 0 5      22.4     1 40    22.4 2  2      0.0
42 2  4 0 5      22.9     1 40    22.9 2  2      0.0
42 2  5 0 5      23.0     1 40    22.5 2  2      0.5
42 2  6 0 4      23.0     1  2     0.5 2 40     22.5
42 2  7 0 3      22.9     1  2     0.0 2 40     22.9
42 2  8 0 3      22.8     1  2     0.0 2 40     22.8
42 2  9 0 5      22.8     1  2     0.0 2 40     22.8
42 2 10 0 4      23.1     1 40    22.6 2  2      0.5

BEST SOLUTION: 111111111111111112121111111111111111111111
```

*Table 12.20*

```
METHOD = 1


42 3  1 0 5  8494.0     1 15 3517.0 2 14 2678.0 3 13 2299.0
42 3  2 0 4  8978.0     1 14 3069.0 2 14 3023.0 3 14 2886.0
42 3  3 0 3  9134.0     1 13 2695.0 2 15 3669.0 3 14 2770.0
42 3  4 0 4  8738.0     1 13 2657.0 2 15 3517.0 3 14 2564.0
42 3  5 0 3  9035.0     1 14 2607.0 2 14 3312.0 3 14 3116.0
42 3  6 0 4  8795.0     1 14 2754.0 2 14 2979.0 3 14 3062.0
42 3  7 0 4  9110.0     1 14 3066.0 2 14 3071.0 3 14 2973.0
42 3  8 0 5  9176.0     1 15 3598.0 2 14 2939.0 3 13 2639.0
42 3  9 0 7  9041.0     1 13 2569.0 2 15 3449.0 3 14 3023.0
42 3 10 0 4  9175.0     1 14 3196.0 2 14 2831.0 3 14 3148.0

BEST SOLUTION: 131122222133111222331331323113223213131212

METHOD = 2

42 3  1 0 8    614.5     1 17   281.1 2 12   136.3 3 13   197.1
42 3  2 0 5    642.9     1 16   241.6 2 13   215.1 3 13   186.2
42 3  3 0 3    632.5     1 17   285.4 2 11   126.6 3 14   220.5
42 3  4 0 2    651.5     1 17   299.3 2 13   195.0 3 12   157.2
42 3  5 0 3    644.4     1 15   241.9 2 16   270.9 3 11   131.6
42 3  6 0 5    618.2     1 12   149.1 2 16   247.4 3 14   221.7
42 3  7 0 3    635.5     1 12   186.8 2 13   165.9 3 17   282.8
42 3  8 0 3    645.0     1 19   335.3 2 11   155.2 3 12   154.5
42 3  9 0 3    649.4     1 13   194.9 2 18   309.0 3 11   145.5
42 3 10 0 6    655.4     1 15   250.4 2 15   236.1 3 12   168.9

BEST SOLUTION: 131123322121311232313311123113221231331212

METHOD = 3

42 3  1 0 5     23.4     1 38    22.9 2  2      0.0 3  2       0.5
42 3  2 0 6     23.0     1  2     0.0 2 38     23.0 3  2       0.0
42 3  3 0 4     22.1     1 38    22.1 2  2      0.0 3  2       0.0
42 3  4 0 3     22.8     1  3     0.2 2  2      0.0 3 37      22.6
42 3  5 0 5     22.7     1  2     0.0 2  2      0.0 3 38      22.7
42 3  6 0 6     22.0     1  2     0.0 2 38     22.0 3  2       0.0
42 3  7 0 3     22.7     1  2     0.0 2  2      0.0 3 38      22.7
42 3  8 0 6     23.0     1 38    22.5 2  2      0.0 3  2       0.5
42 3  9 0 4     23.1     1  2     0.0 2  2      0.5 3 38      22.6
42 3 10 0 3     23.9     1 38    22.9 2  2      0.0 3  2       1.0

BEST SOLUTION: 222222222222222212122322223222222222222222
```

*Table 12.21*

```
METHOD = 1

42 4  1 0 6  5798.0    1  9  819.0 2 11 1778.0 3 11 1544.0 4 11 1657.0
42 4  2 0 5  6092.0    1 11 1578.0 2 11 1629.0 3 10 1598.0 4 10 1287.0
42 4  3 0 4  5675.0    1 10 1195.0 2 11 1687.0 3 11 1715.0 4 10 1078.0
42 4  4 0 7  5403.0    1 10 1060.0 2 11 1417.0 3 11 1457.0 4 10 1469.0
42 4  5 0 6  5643.0    1 11 1415.0 2 10 1216.0 3 11 1678.0 4 10 1334.0
42 4  6 0 4  5996.0    1 10 1285.0 2 11 1798.0 3 10 1438.0 4 11 1475.0
42 4  7 0 6  5748.0    1 10 1304.0 2 10 1274.0 3 11 1482.0 4 11 1688.0
42 4  8 0 5  5647.0    1 10 1135.0 2 12 1964.0 3 10 1323.0 4 10 1225.0
42 4  9 0 5  5867.0    1 10 1445.0 2 11 1362.0 3 11 1596.0 4 10 1464.0
42 4 10 0 6  6003.0    1 10 1112.0 2 11 1911.0 3  9 1188.0 4 12 1792.0

BEST SOLUTION: 222431131232242111441132421423333143444323

METHOD = 2

42 4  1 0 6   531.5    1 13   188.6 2  9   119.7 3 11   129.0 4  9    94.2
42 4  2 0 6   492.3    1 11   126.6 2 12   172.1 3 11   132.5 4  8    61.1
42 4  3 0 3   515.7    1 14   226.4 2  9   104.6 3  8    66.4 4 11   118.5
42 4  4 0 6   502.4    1  7    50.7 2 11   142.4 3 12   149.1 4 12   160.3
42 4  5 0 6   513.8    1 17   276.8 2  9    81.2 3  8    94.8 4  8    61.0
42 4  6 0 3   555.0    1 10    97.7 2 11   162.5 3 11   189.6 4 10   105.2
42 4  7 0 8   494.1    1 11   132.5 2  9    70.4 3 14   208.4 4  8    82.9
42 4  8 0 5   525.9    1 13   199.8 2 10   119.0 3 10   124.0 4  9    83.1
42 4  9 0 3   518.6    1  9    96.3 2 13   180.2 3 12   197.1 4  8    45.0
42 4 10 0 3   555.6    1 11   146.4 2 10   107.1 3 12   195.7 4  9   106.4

BEST SOLUTION: 221134434131211444222231224113333413122323

METHOD = 3

42 4  1 0 4    23.0    1  2     0.0 2  2     0.5 3 36    22.5 4  2     0.0
42 4  2 0 4    22.7    1 36    22.7 2  2     0.0 3  2     0.0 4  2     0.0
42 4  3 0 5    23.0    1 36    22.0 2  2     1.0 3  2     0.0 4  2     0.0
42 4  4 0 4    23.2    1  3     0.2 2  5     0.7 3 30    21.5 4  4     0.8
42 4  5 0 6    22.6    1 36    22.6 2  2     0.0 3  2     0.0 4  2     0.0
42 4  6 0 5    23.1    1  2     0.5 2  2     0.5 3 36    22.1 4  2     0.0
42 4  7 0 4    22.3    1  2     0.0 2  2     0.0 3  2     0.0 4 36    22.3
42 4  8 0 5    24.0    1 36    22.5 2  2     0.0 3  2     1.0 4  2     0.5
42 4  9 0 3    22.9    1  2     0.0 2  2     0.5 3  2     0.0 4 36    22.4
42 4 10 0 5    22.9    1  2     0.0 2  2     0.0 3 36    22.4 4  2     0.5

BEST SOLUTION: 434244144444443414444444444244444444444444
```

*Table 12.22*

```
C   SAMPLE MAIN PROGRAM FOR CLREXM
C
C
C   MDX=200
C   NDX=5
C   SDX=10
C
      DIMENSION A(200,10),B(200),X(5,10),Y(5,10),E(5),
     *          F(10,5),T(10,5),R(5,45),FJ(10),TJ(10),RJ(45),
     *          FP(10),TP(10),RP(45),FQ(10),TQ(10),RQ(45),AI(10)
      INTEGER   S,Z(200),P(200),MJ(5)
      N1=2
      N2=5
      M=96
      S=4
      NRMAX=20
      BIG=1.E50
      KI=5
      KO=6
      WRITE(KO,20)
      DO 1 I=1,M
            READ(KI,26) (A(I,K),K=1,S)
    1 CONTINUE
      READ(KI,26) (B(I),I=1,M)
      DO 2 I=1,M
            WRITE(KO,27) I,B(I),(A(I,K),K=1,S)
    2 CONTINUE
      NS=((S-1)*S)/2
      MO=S
      DO 9 N=N1,N2
            IY=37519281
            WRITE(KO,20)
            DD=BIG
            DO 5 NR=1,NRMAX
                  CALL RANDP (M,N,Z,IY)
                  CALL CLREXM (A,200,M,10,S,B,5,N,NS,Z,MO,MJ,X,
     *                            E,D,IT,IFLAG,AI,F,T,R,
     *                            FJ,TJ,RJ,FP,TP,RP,FQ,TQ,RQ)
                  IF(IFLAG.NE.0) GOTO 5
                  WRITE(KO,22) M,S,N,MO,NR,IT,D,
     *                            (J,MJ(J),E(J),J=1,N)
                  IF(D.GE.DD) GOTO 5
                  DD=D
                  DO 3 I=1,M
                        P(I)=Z(I)
    3             CONTINUE
                  DO 4 K=1,S
                  DO 4 J=1,N
    4             Y(J,K)=X(J,K)
    5       CONTINUE
            WRITE(KO,23) N,NRMAX
            WRITE(KO,25)
            DO 7 J=1,N
                  WRITE(KO,24) J
                  WRITE(KO,25)
                  DO 6 I=1,M
                        IF(J.NE.P(I)) GOTO 6
                        WRITE(KO,27) I,B(I),(A(I,K),K=1,S)
    6             CONTINUE
                  WRITE(KO,25)
    7       CONTINUE
            WRITE(KO,20)
            DO 8 J=1,N
                  WRITE(KO,28) J,(Y(J,K),K=1,S)
```

*Table 12.22 – continued*

```
8       CONTINUE
9 CONTINUE
20 FORMAT('1')
21 FORMAT(' ',I3,3X,5F7.1)
22 FORMAT(' ',I3,3I2,I3,I2,3X,F12.2,5(I2,I3,F12.2))
23 FORMAT('1','BEST SOLUTION FOR',I2,' CLUSTERS AMONG ',I2,
 *        ' TRIALS')
24 FORMAT(' ','CLUSTER ',I1)
25 FORMAT('0')
26 FORMAT(16F5.0)
27 FORMAT(' ',I6,3X,F7.1,3X,5F7.1)
28 FORMAT(' ','COEFFICIENTS IN CLUSTER',I2,' :',3X,5F12.3)
   STOP
   END
```

## 12.4 Sample Main Program and Test Example for Clusterwise Linear Regression

The program for CLREXM shown in Table 12.22 has the usual structure. The values of the independent variables are read into the matrix ((A(I, K), K=1, S), I=1, M), those of the dependent variable into the vector (B(I), I=1, M). The program is set up for a maximum of M = 200, S = 10 and N = 5. Here, it is applied for an example listed in Table 12.23 with M = 96 and S = 4. That table lists values in rows in the form (I, B(I), (A(I, K), K=1, S), I=1, M).

The data for the example were generated as follows. First, clusters of different sizes were chosen: $C_1 = \{1, \ldots, 48\}$, $C_2 = \{49, \ldots, 80\}$ and $C_3 = \{81, \ldots, 96\}$. For these, sets of coefficients were invented. A matrix A was randomly generated, and the scalar products between its $I^{th}$ row and the $Z(I)^{th}$ vector of coefficients ((X(J, K), K=1, S), J=Z(I)) were formed. The resulting M values were altered randomly and by different amounts, and then used as values for the vector (B(I), I=1, M).

Applied to this example, the program needed 362 seconds on the TR 440 to find solutions for N1 = 2 to N2 = 5 with NRMAX = 20 starting partitions each time. To save space, Table 12.24 shows the results for N = 3 only. For NR = 1, NRMAX, each row displays the values of M, S, N, M0, NR, IT, D, (J, MJ(J), E(J), J=1, N). Below, the values of the desired sets of coefficients are printed for NR = 10.

In 18 out of 20 cases the originally given clusters were reproduced with approximately the given coefficients, leading to a value of the objective function of $D \approx 3278$. It is apparent that the rounding error should not be underestimated here, since the 18 values of the objective function, all for the same end partition, already differ in the fifth decimal place. At first sight the values of the objective function for NR = 2 and NR = 14 are astonishingly large. However, this becomes understandable if one takes account of the magnitude of the given data and considers the fact that, for N = 2, the value of D was $\approx 91658789$ NRMAX = 20 times.

Because of this and other experiences with CLREXM concerning sensitivity to rounding errors, we recommend a degree of caution towards the numerical results when larger values of m are involved. If possible, double or multiple precision computations should be used, the minimum number of elements

M0 in each class should be chosen larger than S, and, finally, an end partition found should be used once more as a starting partition. This is easily achieved with a second call to CLREXM.

*Table 12.23*

| | | | | | |
|---|---|---|---|---|---|
| 1 | 4815.0 | 75.0 | 91.0 | 6.0 | 85.0 |
| 2 | 234.0 | 6.0 | 48.0 | 48.0 | 48.0 |
| 3 | 160.0 | 5.0 | 60.0 | 24.0 | 40.0 |
| 4 | 4794.0 | 74.0 | 48.0 | 8.0 | 86.0 |
| 5 | 659.0 | 10.0 | 74.0 | 78.0 | 11.0 |
| 6 | 4551.0 | 67.0 | 89.0 | 32.0 | 19.0 |
| 7 | 3150.0 | 49.0 | 36.0 | 32.0 | 68.0 |
| 8 | 6014.0 | 90.0 | 49.0 | 95.0 | 82.0 |
| 9 | 5017.0 | 74.0 | 90.0 | 9.0 | 19.0 |
| 10 | 6353.0 | 94.0 | 49.0 | 19.0 | 47.0 |
| 11 | 1399.0 | 21.0 | 3.0 | 90.0 | 38.0 |
| 12 | 6550.0 | 95.0 | 21.0 | 60.0 | 34.0 |
| 13 | 606.0 | 14.0 | 50.0 | 7.0 | 84.0 |
| 14 | 6081.0 | 89.0 | 46.0 | 65.0 | 44.0 |
| 15 | 2392.0 | 38.0 | 83.0 | 62.0 | 63.0 |
| 16 | 3194.0 | 48.0 | 67.0 | 33.0 | 33.0 |
| 17 | 4553.0 | 70.0 | 88.0 | 91.0 | 86.0 |
| 18 | 6244.0 | 92.0 | 41.0 | 62.0 | 53.0 |
| 19 | 6379.0 | 92.0 | 3.0 | 4.0 | 17.0 |
| 20 | 2594.0 | 43.0 | 54.0 | 7.0 | 93.0 |
| 21 | 2750.0 | 40.0 | 75.0 | 86.0 | 15.0 |
| 22 | 5024.0 | 74.0 | 64.0 | 72.0 | 39.0 |
| 23 | 3191.0 | 50.0 | 61.0 | 46.0 | 71.0 |
| 24 | 5440.0 | 78.0 | 5.0 | 60.0 | 20.0 |
| 25 | 2014.0 | 32.0 | 13.0 | 14.0 | 56.0 |
| 26 | 1140.0 | 21.0 | 34.0 | 18.0 | 78.0 |
| 27 | 6084.0 | 88.0 | 96.0 | 17.0 | 1.0 |
| 28 | 4773.0 | 71.0 | 4.0 | 74.0 | 67.0 |
| 29 | 2909.0 | 43.0 | 29.0 | 10.0 | 19.0 |
| 30 | 164.0 | 3.0 | 32.0 | 96.0 | 29.0 |
| 31 | 11.0 | 5.0 | 82.0 | 22.0 | 72.0 |
| 32 | 3507.0 | 50.0 | 54.0 | 64.0 | 2.0 |
| 33 | 3107.0 | 48.0 | 33.0 | 54.0 | 67.0 |
| 34 | 2690.0 | 42.0 | 82.0 | 8.0 | 46.0 |
| 35 | 4162.0 | 65.0 | 88.0 | 23.0 | 79.0 |
| 36 | 1898.0 | 32.0 | 33.0 | 35.0 | 86.0 |
| 37 | 2902.0 | 44.0 | 89.0 | 35.0 | 32.0 |
| 38 | 4162.0 | 61.0 | 9.0 | 85.0 | 46.0 |
| 39 | 3427.0 | 52.0 | 14.0 | 62.0 | 65.0 |
| 40 | 3861.0 | 56.0 | 69.0 | 66.0 | 13.0 |
| 41 | 1268.0 | 23.0 | 86.0 | 23.0 | 72.0 |
| 42 | 3549.0 | 54.0 | 64.0 | 24.0 | 47.0 |
| 43 | 5538.0 | 80.0 | 26.0 | 56.0 | 22.0 |
| 44 | 1484.0 | 24.0 | 22.0 | 94.0 | 65.0 |
| 45 | 3363.0 | 50.0 | 87.0 | 32.0 | 21.0 |
| 46 | 4990.0 | 77.0 | 88.0 | 9.0 | 78.0 |
| 47 | 5739.0 | 86.0 | 88.0 | 21.0 | 54.0 |
| 48 | 5287.0 | 79.0 | 47.0 | 66.0 | 65.0 |
| 49 | 3450.0 | 372.0 | 273.0 | 48.0 | 39.0 |
| 50 | 9340.0 | 232.0 | 225.0 | 172.0 | 64.0 |
| 51 | 7806.0 | 160.0 | 162.0 | 146.0 | 31.0 |
| 52 | 5089.0 | 76.0 | 291.0 | 92.0 | 59.0 |
| 53 | 8200.0 | 128.0 | 186.0 | 154.0 | 63.0 |
| 54 | 5980.0 | 224.0 | 30.0 | 110.0 | 3.0 |
| 55 | 3351.0 | 292.0 | 123.0 | 52.0 | 37.0 |
| 56 | 7925.0 | 336.0 | 96.0 | 142.0 | 54.0 |
| 57 | 5030.0 | 388.0 | 255.0 | 78.0 | 96.0 |
| 58 | 6031.0 | 180.0 | 165.0 | 110.0 | 16.0 |
| 59 | 3051.0 | 248.0 | 219.0 | 46.0 | 27.0 |
| 60 | 1969.0 | 88.0 | 60.0 | 34.0 | 40.0 |
| 61 | 5961.0 | 288.0 | 216.0 | 102.0 | 75.0 |
| 62 | 9330.0 | 312.0 | 261.0 | 168.0 | 58.0 |
| 63 | 2433.0 | 12.0 | 141.0 | 44.0 | 66.0 |
| 64 | 4422.0 | 384.0 | 99.0 | 70.0 | 62.0 |

*Table 12.23 – continued*

| | | | | | |
|---|---|---|---|---|---|
| 65 | 5490.0 | 196.0 | 39.0 | 100.0 | 56.0 |
| 66 | 10155.0 | 196.0 | 183.0 | 190.0 | 88.0 |
| 67 | 6972.0 | 16.0 | 30.0 | 138.0 | 18.0 |
| 68 | 2341.0 | 376.0 | 114.0 | 28.0 | 76.0 |
| 69 | 662.0 | 152.0 | 183.0 | 2.0 | 64.0 |
| 70 | 10302.0 | 276.0 | 297.0 | 188.0 | 48.0 |
| 71 | 3524.0 | 392.0 | 3.0 | 54.0 | 42.0 |
| 72 | 5117.0 | 344.0 | 273.0 | 82.0 | 48.0 |
| 73 | 3057.0 | 220.0 | 105.0 | 50.0 | 1.0 |
| 74 | 6589.0 | 388.0 | 144.0 | 112.0 | 60.0 |
| 75 | 1005.0 | 56.0 | 33.0 | 16.0 | 73.0 |
| 76 | 6568.0 | 160.0 | 192.0 | 120.0 | 60.0 |
| 77 | 4790.0 | 128.0 | 87.0 | 88.0 | 59.0 |
| 78 | 7151.0 | 164.0 | 132.0 | 132.0 | 80.0 |
| 79 | 982.0 | 232.0 | 45.0 | 8.0 | 77.0 |
| 80 | 2745.0 | 132.0 | 174.0 | 46.0 | 2.0 |
| 81 | 3691.0 | 138.0 | 55.0 | 16.0 | 15.0 |
| 82 | 4284.0 | 66.0 | 95.0 | 12.0 | 57.0 |
| 83 | 1981.0 | 6.0 | 50.0 | 44.0 | 36.0 |
| 84 | 2859.0 | 36.0 | 70.0 | 96.0 | 51.0 |
| 85 | 4105.0 | 60.0 | 100.0 | 64.0 | 69.0 |
| 86 | 3544.0 | 36.0 | 55.0 | 0.0 | 15.0 |
| 87 | 7278.0 | 138.0 | 120.0 | 20.0 | 42.0 |
| 88 | 8697.0 | 6.0 | 115.0 | 72.0 | 12.0 |
| 89 | 2625.0 | 126.0 | 85.0 | 4.0 | 72.0 |
| 90 | -3326.0 | 36.0 | 5.0 | 64.0 | 66.0 |
| 91 | 5197.0 | 132.0 | 110.0 | 16.0 | 63.0 |
| 92 | -51.0 | 36.0 | 0.0 | 88.0 | 6.0 |
| 93 | 5185.0 | 42.0 | 70.0 | 28.0 | 9.0 |
| 94 | 2823.0 | 24.0 | 85.0 | 48.0 | 69.0 |
| 95 | 3697.0 | 60.0 | 45.0 | 72.0 | 3.0 |
| 96 | 65.0 | 132.0 | 35.0 | 4.0 | 48.0 |

*Table 12.24*

| | | | | | | | | | | | | | | | |
|---|---|---|---|---|---|---|---|---|---|---|---|---|---|---|---|
| 96 | 4 | 3 | 4 | 1 | 5 | 3277.76 | 1 | 16 | 280.48 | 2 | 32 | 1664.53 | 3 | 48 | 1332.71 |
| 96 | 4 | 3 | 4 | 2 | 8 | 35317391.50 | 1 | 44 | 1515628.97 | 2 | 36 | 30731786.06 | 3 | 16 | 3069976.44 |
| 96 | 4 | 3 | 4 | 3 | 5 | 3278.42 | 1 | 16 | 280.59 | 2 | 48 | 1333.25 | 3 | 32 | 1664.58 |
| 96 | 4 | 3 | 4 | 4 | 4 | 3277.74 | 1 | 32 | 1664.60 | 2 | 48 | 1332.48 | 3 | 16 | 280.60 |
| 96 | 4 | 3 | 4 | 5 | 3 | 3278.12 | 1 | 48 | 1333.08 | 2 | 16 | 280.49 | 3 | 32 | 1664.56 |
| 96 | 4 | 3 | 4 | 6 | 5 | 3277.67 | 1 | 16 | 280.53 | 2 | 32 | 1664.55 | 3 | 48 | 1332.55 |
| 96 | 4 | 3 | 4 | 7 | 5 | 3277.59 | 1 | 16 | 280.74 | 2 | 32 | 1664.60 | 3 | 48 | 1332.19 |
| 96 | 4 | 3 | 4 | 8 | 8 | 3277.55 | 1 | 32 | 1664.57 | 2 | 48 | 1332.31 | 3 | 16 | 280.63 |
| 96 | 4 | 3 | 4 | 9 | 4 | 3277.55 | 1 | 16 | 280.60 | 2 | 32 | 1664.57 | 3 | 48 | 1332.41 |
| 96 | 4 | 3 | 4 | 10 | 5 | 3277.43 | 1 | 32 | 1664.58 | 2 | 16 | 280.41 | 3 | 48 | 1332.44 |
| 96 | 4 | 3 | 4 | 11 | 5 | 3277.50 | 1 | 48 | 1332.31 | 2 | 32 | 1664.56 | 3 | 16 | 280.61 |
| 96 | 4 | 3 | 4 | 12 | 10 | 3278.11 | 1 | 32 | 1664.58 | 2 | 48 | 1332.86 | 3 | 16 | 280.60 |
| 96 | 4 | 3 | 4 | 13 | 5 | 3277.84 | 1 | 16 | 280.60 | 2 | 32 | 1664.64 | 3 | 48 | 1332.55 |
| 96 | 4 | 3 | 4 | 14 | 7 | 94160789.67 | 1 | 42 | 49724482.01 | 2 | 26 | 17819783.36 | 3 | 28 | 26616524.21 |
| 96 | 4 | 3 | 4 | 15 | 6 | 3277.89 | 1 | 32 | 1664.48 | 2 | 16 | 280.33 | 3 | 48 | 1332.95 |
| 96 | 4 | 3 | 4 | 16 | 5 | 3277.86 | 1 | 16 | 280.50 | 2 | 48 | 1332.69 | 3 | 32 | 1664.61 |
| 96 | 4 | 3 | 4 | 17 | 4 | 3277.76 | 1 | 48 | 1332.56 | 2 | 32 | 1664.57 | 3 | 16 | 280.55 |
| 96 | 4 | 3 | 4 | 18 | 3 | 3277.94 | 1 | 16 | 280.66 | 2 | 48 | 1332.68 | 3 | 32 | 1664.57 |
| 96 | 4 | 3 | 4 | 19 | 4 | 3277.61 | 1 | 48 | 1332.48 | 2 | 16 | 280.54 | 3 | 32 | 1664.58 |
| 96 | 4 | 3 | 4 | 20 | 4 | 3277.75 | 1 | 16 | 280.40 | 2 | 32 | 1664.57 | 3 | 48 | 1332.77 |

| | | | | |
|---|---|---|---|---|
| COEFFICIENTS IN CLUSTER 1 : | 2.017 | 1.011 | 49.959 | 0.936 |
| COEFFICIENTS IN CLUSTER 2 : | 1.054 | 79.941 | 3.018 | -60.002 |
| COEFFICIENTS IN CLUSTER 3 : | 69.958 | -0.968 | 1.004 | -4.017 |

**Problem 12.1**: Perform the conversion of OVSEXM and OVREXM to handle real data matrices, as described in Section 10.1. Will you need to add more than one statement to the (correspondingly modified) main program if you first wish to process the random starting partitions with TRWEXM and take the resulting end partitions as starting partitions for OVSEXM or OVREXM in their modified form? Does this procedure save computer time? Use the data matrices listed in Tables 12.1 and 12.7 (after adding + 1) as well as the first four examples Fig. 11.1.

**Problem 12.2** (Continuation of Problem 10.3): Apply the subroutine you have written to the data matrix or Table 12.1, compute using BVPEXM, and compare the computation time required with that needed by OVPEXM with the untransformed data matrix.

**Problem 12.3**: Treat the data matrix in Table 12.7 on your computer, using BVPEXM and OVPEXM in turn (again, +1 must be added beforehand). Compare the times required.

**Problem 12.4** (Continuation of Problem 10.5): For the data matrices of Tables 12.1 and 12.7 use the Jaccard metric to calculate distances, and then apply TIHEXM. (With Table 12.1, do you need to subtract +1 first?) How do the results for METHOD = 1, 2, 3 differ from the possibly optimal partitions in Table 12.4 and in Tables 12.10 and 12.12 when the $L_1$-criterion is used?

**Problem 12.5**: Make up a test data generator for the problem of clusterwise linear regression, and use it to test CLREXM.

**Problem 12.6**: How would you generate random starting partitions if you wanted to apply CLREXM to data such as Fig. 8.1?

**Problem 12.7** (Continuation of Problem 11.6): Perform corresponding investigations using the main programs in this chapter.

# Appendix

# Description of the magnetic tape with all programs

The twelve sample main programs together with all the subroutines they call and with the corresponding test data are stored on a magnetic tape. As a rule, all twelve programs will run immediately without alterations or additions.

The tape has been written with NOLABEL, EBCDIC, 1600 BPI. It has a block structure of LRECL = 80 and BLKSIZE = 800. Thus, it should be possible to read it in every computer centre.

Corresponding to the twelve main programs, the tape contains twelve files with 362, 343, 403, 489, 502, 622, 589, 502, 255, 243, 241 and 405 card images respectively (LRECL = 80). Table 13.1 indicates which main programs (x) use which subroutines (y) and (z). Every file contains all subroutines called in it.

The first five files contain the main programs for TRWEXM, TRWMDM, the combination TRWMDM/TRWEXM, DETEXM and DWBEXM. The sets of the frequently mentioned twelve examples, the first eight of which are given in Chapter 11.

Files 6 to 8 contain the programs used to test OVSEXM, OVREXM and OVPEXM, with a set of test data that is the same in all three cases. It is larger than that listed in Table 12.1. It comprises S = 260 variables, the first 122 of which coincide with the Table. Note the FORMAT (3X, 24I1). After the test data, a few statements are added. These would need to be exchanged in the main programs if one was to use the data set with M = 260 and S = 24.

File 9 comprises the main program in Table 12.8, the subroutine BVPEXM, and the data in Table 12.7 which are to be read with FORMAT (20X, 46I1). Files 10 and 11 correspond to the two examples of applying TIHEXM. Finally, file 12 contains the sample main program for CLREXM and also comprises the associated set of test data listed in Table 12.23.

Before using the subroutines for his own applications, the user is advised to fetch the relevant file from the tape and try to reproduce the results for the given examples. Since all main programs use URAND, this is possible, however, only with the provisos described in Section 11.1.

All enquiries concerning the availability of this type should be addressed to the publisher:

Ellis Horwood Ltd., Market Cross House,
Chichester, England, PO19 1EB

*Table 13.1*

| Main programs | 1 | 2 | 3 | 4 | 5 | 6 | 7 | 8 | 9 | 10 | 11 | 12 |
|---|---|---|---|---|---|---|---|---|---|---|---|---|
| Subroutines | | | | | | | | | | | | |
| TRWEXM | x | | x | | | | | | | | | |
| TRWMDM | | x | x | | | | | | | | | |
| DETEXM | | | | x | | | | | | | | |
| DWBEXM | | | | | x | | | | | | | |
| OVSEXM | | | | | | x | | | | | | |
| OVREXM | | | | | | | x | | | | | |
| OVPEXM | | | | | | | | x | | | | |
| BVPEXM | | | | | | | | | x | | | |
| TIHEXM | | | | | | | | | | x | x | |
| CLREXM | | | | | | | | | | | | x |
| MEANS | y | y | y | y | y | | | | | | | |
| TRACES | y | y | y | | | | | | | | | |
| LDLT | | | | y | y | | | | | | | |
| UPDATE | | | | y | y | | | | | | | |
| DISTW | | | | y | | | | | | | | |
| WJSCAT | | | | | y | | | | | | | |
| MEDIAN | | | | | | | | y | | | | |
| INEXCL | | | | | | | | | | | | y |
| RANDP | z | z | z | z | z | z | z | z | z | z | z | z |
| URAND | z | z | z | z | z | z | z | z | z | z | z | z |
| PRINT | z | z | z | z | z | | | | | | | |
| TRAFOR | | | | z | z | | | | | | | |

## Hints for Implementation on a Microcomputer

The following remarks refer to Microsoft FORTRAN-80, which is available for microcomputers equipped with 8080 or Z80 processors. Note that REAL magnitudes lie in the range from $10^{-38}$ to $10^{38}$ approximately and have a precision about 7 decimal places. INTEGERS are restricted to the range from −32768 to 32767 and LOGICAL variables are stored in one byte, corresponding to LOGICAL*1. Where a larger integer range is required, an INTEGER*4 declaration is possible, allowing numbers between −2147 483 648 and 2147 483 647.

Correspondingly, the subroutines TRWMDM, TRWEXM, DETEXM, DWBEXM, TIHEXM and CLREXM (and routines called by these) can run immediately, provided that the statement BIG = 1.E50 is changed to BIG = 1.E35. There should be no need to modify EPS1, EPS2 and EPS3.

In OVSEXM, OVREXM and OVPEXM, the data matrix X (and XX) can remain INTEGER for most applications. However, KBIG or BIG, and quantities such as D, F, EP, EJ, EQ, and for the array E have to be declared INTEGER*4 if it is to be expected that the INTEGER range described above might be exceeded. Unfortunately, X and XX also need to be declared as INTEGER*4 in this case. KBIG and BIG must be changed to 32767 (INTEGER*2) or 2147 483 647 (INTEGER*4).

Finally, in BVPEXM, the LOGICAL declaration of X already implies that the space-saving one-byte version is chosen. But, for safety, BIG, D, E, F, EJ, EP, EQ and E should again be declared as INTEGER*4.

With these minor modifications, the subroutines should run. In testing the quoted main programs, we must, however, note that URAND does not work. Instead, the following generator (B. A. Wichmann, I. D. Hill: Algorithm AS183: An efficient and portable pseudo-random number generator, *Applied Statistics* **31**, 188–190 (1982)) could be used. This may also be used on a mainframe if the conditions for using URAND as in Table 11.2 are not satisfied (see Section 11.2):

```
  FUNCTION URAND(L)
  COMMON /RAND/ IX,IY,IZ
  IX=171*MOD(IX,177)-2*(IX/177)
  IY=172*MOD(IY,176)-35*(IY/176)
  IZ=170*MOD(IZ,178)-63*(IZ/178)
  IF (IX.LT.0) IX=IX+30269
  IF (IY.LT.0) IY=IY+30307
  IF (IZ.LT.0) IZ=IZ+30323
  URAND=AMOD(FLOAT(IX)/30269.0+FLOAT(IY)/30307.0
1          +FLOAT(IZ)/30323.0, 1.0)
  RETURN
  END
```

This subroutine is adjusted to the size of integer available. In the main programs it requires the declaration

```
COMMON /RAND/ IX,IY,IZ
```

and three initialization statements (instead of one for IY), e.g.

```
IX=11
IY=111
IZ=1111
```

In the main programs, KI and KO, the input and output unit numbers, may also need to be changed if required, array dimensions may need to be reduced, as well as associated quantities such as MDX.

As regards computation time, first experiments suggest that on Alphatronic P2S under the CP/M operating system with 48K bytes of working memory is about 100 times slower than the TR 440. This should not prove a limitation with a personal computer.

Readers who are interested in an implementation on a microcomputer – especially on IBM PC XT – should approach the author, since a program package is being developed which uses an interactive file management program to call up the subroutines quoted as well as a number of others. The package is characterized by the following three figures, and might also be implemented on a mainframe.

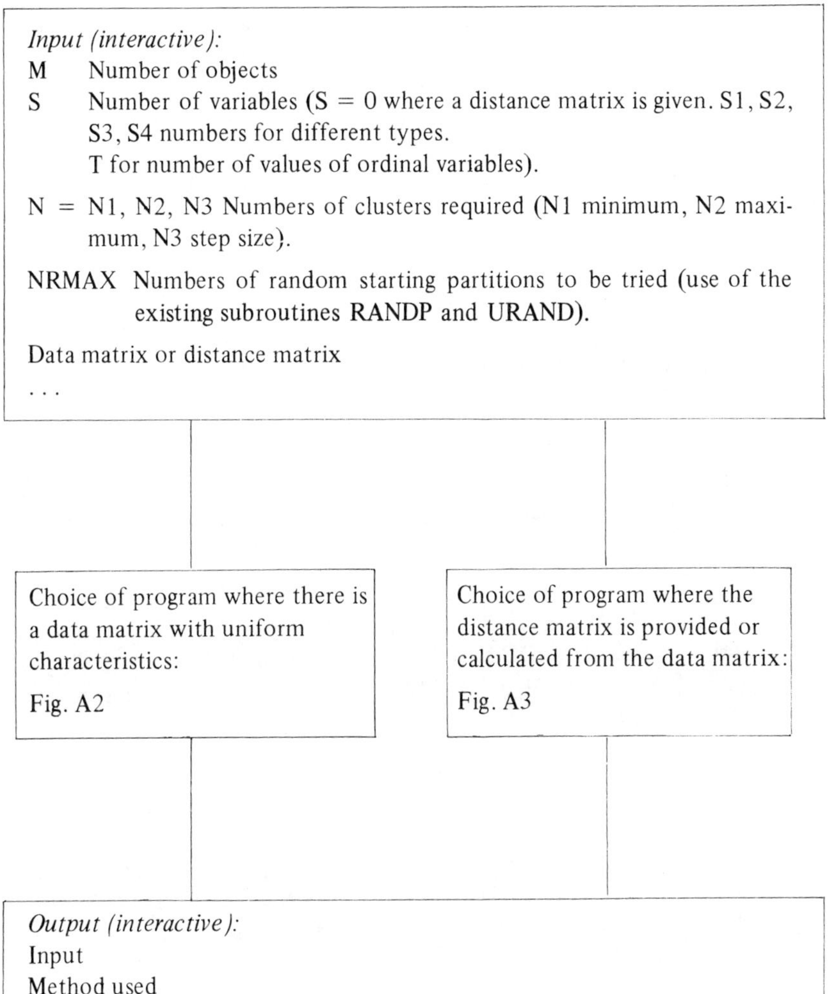

*Input (interactive):*
M Number of objects
S Number of variables (S = 0 where a distance matrix is given. S1, S2, S3, S4 numbers for different types.
T for number of values of ordinal variables).
N = N1, N2, N3 Numbers of clusters required (N1 minimum, N2 maximum, N3 step size).
NRMAX Numbers of random starting partitions to be tried (use of the existing subroutines RANDP and URAND).
Data matrix or distance matrix
. . .

Choice of program where there is a data matrix with uniform characteristics:
Fig. A2

Choice of program where the distance matrix is provided or calculated from the data matrix:
Fig. A3

*Output (interactive):*
Input
Method used
Centres and their profiles
Cluster members ordered by their distance from the cluster centre
Cluster homogeneities, value of the objective function
. . .

Fig. A1 – Structure of the proposed program packages.

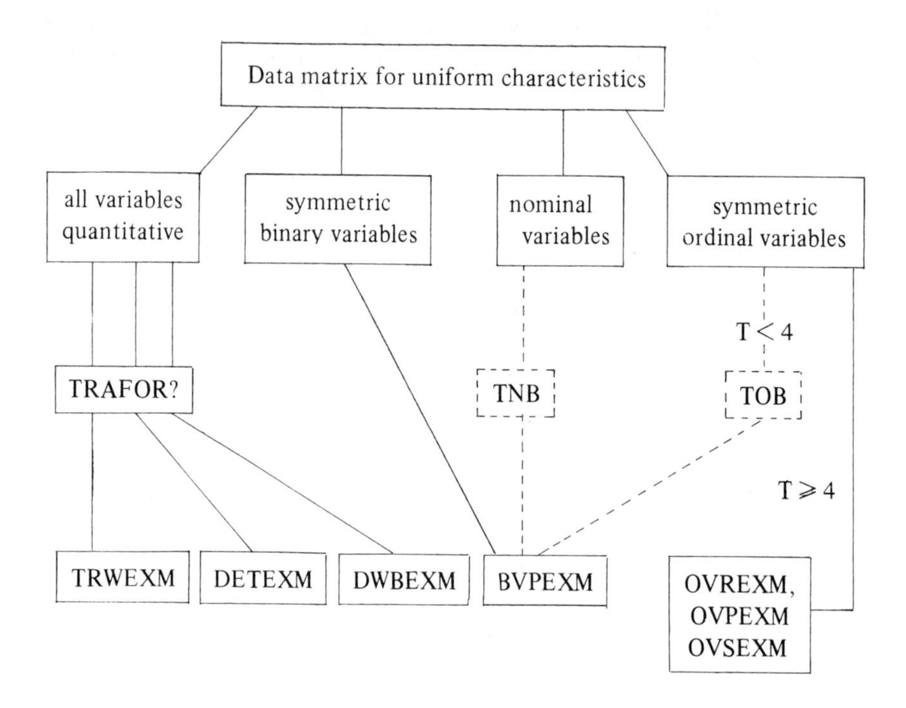


Fig. A2 – Choice of program where a data matrix with uniform characteristics exists.

The subroutines TNB and TOB within Fig. A2 and the subroutines DLP, DJB, DJO, and DADD within Fig. A3 have just to be written. They all will have only few statements and partly are the rusults of problems posed in parts II and III. Their task is

TNB: Transformation of nominal to binary variables
TOB: Transformation of ordinal to binary variables (c.f. (10.6))
DLP: Calculation of $L_p$ distance values
DJB: Calculation of Jaccard metric for binary vectors (c.f. (7.3))
DJO: Calculation of Jaccard metric for ordinal vectors (c.f. (7.3))
DADD: Addition of different distances for partial vectors (c.f. (7.4)).

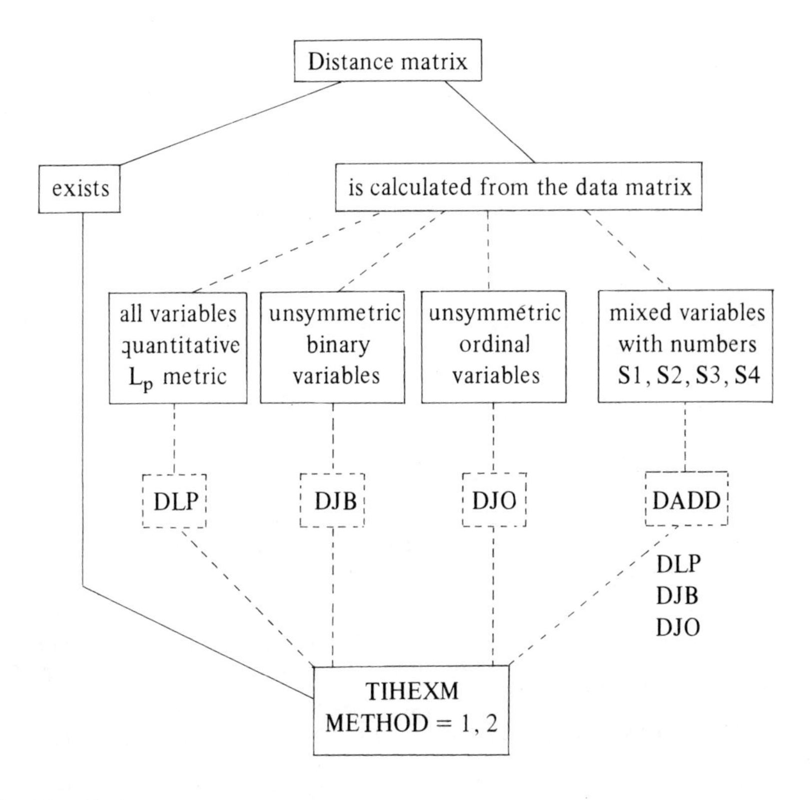


Fig. A3 – Choice of program when using a given distance matrix or calculating it from the data matrix.

# Bibliography


## A Important monographs and survey articles

[A1] *Anderberg, M. R.:* Cluster Analysis for Applications, Academic Press, New York 1973.

[A2] *Bezdek, J.:* Pattern Recognition with Fuzzy Objective Functions, Plenum, New York 1981.

[A3] *Bijnen, E. J.:* Cluster Analysis – Survey and Evaluation of Techniques, Tilburg University Press, The Netherlands 1973.

[A4] *Bock, H. H.:* Automatische Klassifikation, Vandenhoeck & Ruprecht, Göttingen 1974.

[A5] *Bock, H. H.:* Clusteranalyse – Überblick und neuere Entwicklungen, OR Spektrum 1, 211–232 (1980).

[A6] *Cormack, R. M.:* A Review of Classification, J. R. Stat. Soc. A 134, 321–367 (1971).

[A7] *Diday, E., Simon, J. C.:* Clustering Analysis, in: FU, K. S. (Ed.): Digital Pattern Recognition, Springer, New York 1976.

[A8] *Duran, B. S., Odell, P. L.:* Cluster Analysis – A Survey, Springer, Berlin 1974.

[A9] *Everitt, B.:* Cluster Analysis, Heinemann, London 1974.

[A10] *Gordon, A. D.:* Classification, Chapman and Hall, London 1981.

[A11] *Hartigan, J.:* Clustering Algorithms, Wiley, New York 1975.

[A12] *Jambu, M.:* Classification automatique pour l'analyse des données, Vol. 1, 2, Dunod, Paris 1978.

[A13] *Jardine, N., Sibson, R.:* Mathematical Taxonomy, Wiley, New York 1971.

[A14] *Lerman, I. C.:* Les bases de la classification automatique, Gauthier-Villars, Paris 1970.

[A15] *Marcotorchino, J.-F., Michaud, P.:* Optimisation en Analyse Ordinale des Données, Masson, Paris 1980.

[A16] *Opitz, O.:* Numerische Taxonomie, Fischer, Stuttgart 1980.

[A17] *Schader, M.:* Scharfe und unscharfe Klassifikation qualitativer Daten, Verlagsgruppe Athenäum, Hain, Scriptor, Hanstein, Königsstein/Ts. 1981.

[A18] *Sneath, P. H. A., Sokal, R. R.:* Numerical Taxonomy, Freeman & Co., San Francisco 1973.

[A19] *Sodeur, W.:* Empirische Verfahren zur Klassifikation, Teubner, Stuttgart 1974.

[A20] *Späth, H.:* Cluster Analysis Algorithms, Horwood, Chichester 1980. 2nd Edition 1982. (German original: R. Oldenbourg, Munich 1975, 2nd edition 1977.)

[A21] *Späth, H.* (Hrsgb.): Fallstudien Cluster-Analyse, R. Oldenbourg, München 1977.

[A22] *Steinhausen, D., Langner, K.:* Clusteranalyse, de Gruyter, Berlin 1977.

[A23] *Vogel, F.:* Probleme und Verfahren der Numerischen Klassifikation, Vandenhoeck & Ruprecht, Göttingen 1975.

## B References with a direct bearing on the material presented in this book

[B1] *Abdelmalek, N. N.:* Algorithm 551: A Fortran Subroutine for the $L_1$ Solution of Overdetermined Systems of Linear Equations, ACM Trans. Math. Software 6, 228–230 (1980).

[B2] *Anderson, T. W.:* An Introduction to Multivariate Statistical Analysis, Wiley, New York 1958.

[B3] *Arthanari, T. S., Dodge, Y.:* Mathematical Programming in Statistics, Wiley, New York 1981.

[B4] *Banfield, C. F., Bassill, L. C.:* Algorithm AS 113: A Transfer Algorithm for Non-hierarchical Classification, Appl. Stat. 26, 206–210 (1977).

[B5] *Bard, Y.:* Nonlinear Parameter Estimation, Academic Press, New York 1974.

[B6] *Barrodale, I., Roberts, F. D. K.:* Algorithm 478: Solution of an Overdetermined System of Equations in the $L_1$ Norm, Comm. ACM 17, 319–320 (1974).

[B7] *Belschner, W., Späth, H.:* Versuch einer Kategorisierung von erzieherischen Situationsdefinitionen mittels Cluster-Analyse, Psychol. in Erziehung und Unterricht 24, 49–53 (1977).

[B8] *Braun, H.:* Strukturanalyse eines Tankstellennetzes, in [B43, 9–29].

[B9] *Chatfield, C., Collins, A. J.:* Introduction to Multivariate Analysis, Chapman and Hall, London 1980.

[B10] *Clarke, M. R. B.:* Algorithm AS 163: A Givens Algorithm for Moving from one Linear Model to another without going back to the Data, Appl. Stat. 30, 198–203 (1981).
[B11] *Daniel, J. W., Gragg, W. B., Kaufmann, L., Stewart, G. W.:* Reorthogonalization and Stable Algorithms for Updating the Gram-Schmidt QR Factorization, Math. Comp. 30, 772–795 (1976).
[B12] *Diday, E., Govaert, G.:* Classification avec distance adaptive, C. R. Acad. Sci. Pari 278 A, 993–995 (1974).
[B13] *Diday, E., Govaert, G.:* Classification automatique avec distances adaptives, Rev. Fr. Automat. Inf. Rech. Operat. Inf./Comput. Sci. 11, 329–349 (1977).
[B14] *Diday, E., Schroeder, A.:* A New Approach in Mixed Distribution Detection, Rev. Fr. Automat. Inf. Rech. Operat. 10, 75–106 (1976).
[B15] *Dobbener, R.:* Zur Skalen- und Translationsinvarianz von Metriken, Internat. Classification 8, 64–68 (1981).
[B16] *Duda, R. O., Hart, P. E.:* Pattern Classification and Scene Analysis, Wiley, New York 1973.
[B17] *Forsythe, G. E., Malcolm, M. A., Moler, C. B.:* Computer Methods for Mathematical Computations, Prentice Hall, Englewood Cliffs 1977.
[B18] *Friedman, H. P., Rubin, J.:* On Some Invariant Criteria for Grouping Data, J. Am. Stat. Assoc. 62, 1159–1178 (1967).
[B19] *Gentleman, W. M.:* Least Squares Computations by Givens Transformations Without Square Roots, J. Inst. Maths Applics 12, 329–336 (1973).
[B20] *Gentleman, M. W.:* Algorithm AS 75: Basic Procedures for Large, Sparse or Weighted Linear Least Squares Problems, Appl. Stat. 23, 448–454 (1974).
[B21] *Gill, P. E., Golub, G. H., Murray, W., Saunders, M. A.:* Methods for Modifying Matrix Factorizations, Math. Comp. 28, 505–535 (1974).
[B22] *Gill, P. E., Murray, W., Saunders, M. A.:* Methods for Computing and Modifying the LDV Factors of a Matrix, Math. Comp. 29, 1051–1077 (1975).
[B23] *Gordon, A. D., Henderson, J. T.:* An Algorithm for Euclidean Sum of Squares Classification, Biometrics 33, 355–362 (1977).
[B24] *Hand, D. J.:* Discrimination and Classification, Wiley, Chichester 1981.
[B25] *Hartigan, J. A., Wong, M. A.:* Algorithm AS 136: A K-Means Clustering Algorithm, Appl. Stat. 28, 100–108 (1979).
[B26] *Herden, W., Steinhausen, D.:* Ein Verfahren zur Berechnung eines globalen Minimums beim Varianzkriterium in der Cluster-Analyse, in [B45, 115–142].
[B27] *Kernighan, B. W., Lin, S.:* An Efficient Heuristic Procedure for Partitioning Graphs, Bell Systems Techn. J. 49, 291–307 (1970).
[B28] *Koontz, W. L. G., Narenda, P. M., Fukunaga, K.:* A Branch and Bound Clustering Algorithm, IEEE Trans. Comp. 24, 908–914 (1975).
[B29] *Lawson, C. L., Hanson, R. J.:* Solving Least Squares Problems, Prentice Hall, Englewood Cliffs 1974.
[B30] *Marcus, M., Minc, H.:* A Survey of Matrix Theory and Matrix Inequalities, Allyn and Bacon, Boston 1964.
[B31] *Maronna, R., Jakovkis, P. M.:* Multivariate Clustering Procedures with Variable Metrics, Biometrics 30, 499–505 (1974).
[B32] *Marriot, F. H. C.:* The Interpretation of Multiple Observations, Academic Press, London 1974.
[B33] *Merle, G., Späth, H.:* Computational Experiences with Discrete $L_p$-Approximation, Computing 12, 315–321 (1974).
[B34] *Mitrinovic, D. S.:* Analytic Inequalities, Springer, Berlin 1970.
[B35] *Müller, R.:* Untersuchungen zur Methode der adaptiven Distanzen in der Cluster-Analyse, Diplomarbeit, Universität Oldenburg 1979.
[B36] *Peters, U., Willms, C.:* Up- and Downdating Procedures for Linear $L_1$ Regression, OR Spektrum 5, 229–239 (1983).
[B37] *Sherman, J., Morrison, W. J.:* Adjustment of an Inverse Matrix corresponding to Changes in the Elements of a Given Column or a Given Row of the Original Matrix, Ann. Math. Stat. 20, 621 (1940).
[B38] *Späth, H.:* Algorithmen für multivariable Ausgleichsmodelle, R. Oldenbourg, München 1974.
[B39] *Späth, H.:* Algorithm 30: $L_1$ Cluster Analysis, Computing 16, 379–387 (1976).
[B40] Späth, H.: Numerischer Vergleich von zwei kanonischen Varianten des Austauschverfahrens beim Varianzkriterium in der Cluster-Analyse, Angew. Inf. 9, 395–397 (1977).

[B41] *Späth, H.:* Partitionierende Cluster-Analyse bei Binärdaten am Beispiel von bundesdeutschen Hochschulen und Diplomstudiengängen, Z. für Oper. Res. 21, 85–96 (1977).

[B42] *Späth, H.:* Partitionierende Cluster-Analyse für große Objektemengen mit binären Merkmalen am Beispiel von Firmen und deren Berufsgruppenbedarf, in [A21, 63–80].

[B43] *Späth, H.* (Ed.): Fallstudien Operations Research, Band 1, R. Oldenbourg, München 1978.

[B44] *Späth, H.:* Bedarfsvorhersage für saisonale Großsortimente in Handelsunternehmen, in [B43, 134–149].

[B45] *Späth, H.* (Ed.): Ausgewählte Operations Research Software in FORTRAN, R. Oldenbourg, München 1979.

[B46] *Späth, H., Müller, R.:* Das Austauschverfahren für die skalierungsinvariante Methode der adaptiven Distanzen in der Cluster-Analyse, in [B45, 143–163].

[B47] *Späth, H.:* Klassenweise diskrete Approximation, in *Collatz, L., Meinardus, G., Wetterling, W.* (Eds.): Numerische Methoden bei graphentheoretischen und kombinatorischen Problemen, Band 2, ISNM Vol. 46, Birkhäuser, Basel 1979.

[B48] *Späth, H.:* Algorithm 39: Clusterwise Linear Regression, Computing 22, 367–373 (1979).

[B49] *Späth, H.:* Correction to Algorithm 39: Clusterwise Linear Regression, Computing 26, 275 (1981).

[B50] *Späth, H.:* Algorithm 48: A Fast Algorithm for Clusterwise Linear Regression, Computing 29, 175–181 (1982).

[B51] *Späth, H.:* The Minisum Location Problem for the Jaccard Metric, OR Spektrum 3, 91–94 (1981).

[B52] *Steinhausen, D., Steinhausen, J.:* Cluster-Analyse als Instrument der Zielgruppendefinition in der Marktforschung, in [A21, 9–36].

[B53] *Stewart, G. W.:* The Effects of Rounding Error on an Algorithm for Downdating a Cholesky Factorization, J. Inst. Maths Applics 23, 203–213 (1979).

[B54] *Stoer, J.:* Einführung in die Numerische Mathematik I, Springer, Berlin 1976.

[B55] *Streit, U.:* Kombinatorische Optimierung mit Nebenbedingungen am Beispiel des Varianzkriteriums in der Cluster-Analyse, Diplomarbeit, Universität Oldenburg 1979.

[B56] *Watson, G. A.:* On Two Methods for Discrete $L_p$-Approximation, Computing 18, 263–266 (1977).

[B57] *Wolfe, J. M.:* On the Convergence of an Algorithm for a Discrete $L_p$-Approximation, Numer. Math. 32, 439–459 (1979).

[B58] *Watson, G. A.:* An Algorithm for the Single Facility Location Problem Using the Jaccard Metric, SIAM J. Sci. Stat. Comput. 4, 748–756 (1983).

# Index of Symbols

# Index